The Book of
Constellations

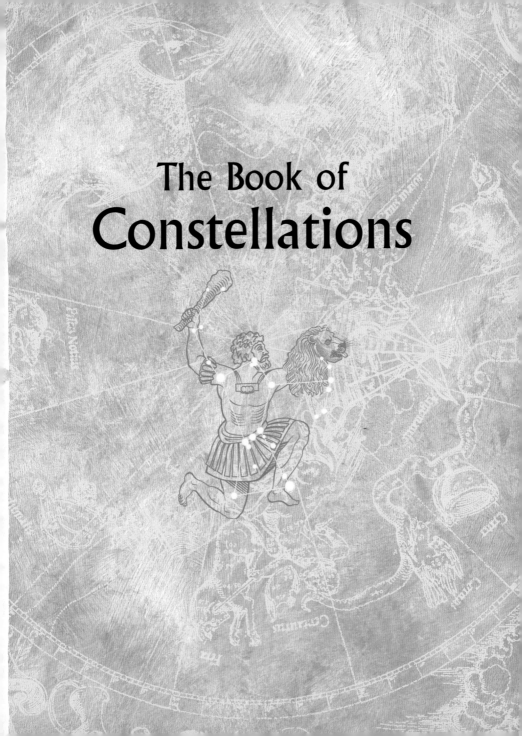

The Book of
Constellations
DISCOVER THE SECRETS IN THE STARS

Robin Kerrod

BARRON'S

A QUARTO BOOK

First edition for the United States, its territories and
dependencies and Canada, published in 2002 by
Barron's Educational Series, Inc.

All inquiries to be addressed to:
Barron's Educational Series, Inc.
250 Wireless Boulevard
Hauppauge, NY 11788
http://www.barronseduc.com

Library of Congress Catalog Card Number
2001087444

International Standard Book Number 0-7641-5440-0

QUAR.BCON

Conceived, designed, and produced by
Quarto Publishing plc
The Old Brewery
6 Blundell Street
London N7 9BH

Senior project editor Nicolette Linton
Senior art editor Sally Bond
Assistant art director Penny Cobb
Text editors Jean Coppendale, Paula Regan
Designer Trevor Newman
Illustrators John Woodcock, Sally Bond
Picture research Image Select International
Indexer Pamela Ellis

Art director Moira Clinch
Publisher Piers Spence

Manufactured by Universal Graphics Pte Ltd, Singapore
Printed by Midas Printing Ltd, China

9 8 7 6 5 4 3 2 1

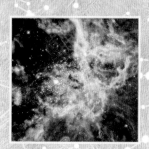

Contents

INTRODUCTION
Dome of the Heavens

When you go stargazing on a clear, dark night, the star-studded heavens seem to form a vast, dark dome over your head. It is the same everywhere on Earth. Our planet seems surrounded by a limitless celestial sphere, and the stars appear to be fixed to the inside of this sphere. As time goes by, the stars seem to wheel over your head and the sphere then seems to be spinning around the Earth.

The appearance of Halley's Comet in 1546 scared the wits out of people.

Gas clouds in the Eagle Nebula, where stars are being born.

At first glance, there is little to distinguish individual stars, and they seem to be scattered haphazardly around the dark dome of the heavens. But it soon becomes evident that the stars are not all alike. Some are so bright they stand out like beacons; others are so dim you can scarcely make them out.

In your mind's eye, you can group together some of the bright stars to make patterns. And night after night, you will be able to find these same patterns in the sky. Even though the stars appear to wheel overhead every night, they move together bodily—they do not change their relative positions in the sky, in their patterns, which we call the constellations.

The stars seem to be fixed in position inside the celestial sphere. This is why they are often called the fixed stars. But there seem to be a few notable exceptions

The constellations of the northern hemisphere, as pictured by Cellarius (1708).

to this general observation: occasionally, five bright objects can be found wandering around the celestial sphere among the fixed stars in the constellations.

But appearances can be deceiving. We know that there is no great enveloping dark celestial sphere surrounding the Earth. The darkness of the night sky is the profound blackness of empty space extending for distances so vast as to be beyond our human comprehension. The tiny pinpricks of light visible as stars are, in reality, huge globes of incandescent gas that pour forth enormous energy into space as light, heat, and other forms of radiation. They are distant Suns. As for the wandering objects, they are not stars at all, but much smaller, closer bodies that we call the planets.

The Romans named the planet Mars after their god of war, whom the Greeks called Ares.

Aquarius

Ancient astronomers saw mythological figures in the star patterns of the constellations.

Chapter 1·THE UNIVERSE
but not as we know it

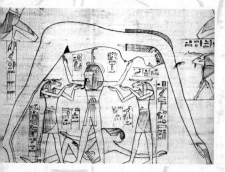

In ancient Egyptian mythology, the starry body of the sky goddess, Nut, formed the heavens.

Two thousand years ago, astronomers believed that there was a celestial sphere surrounding Earth, and that Earth was the center of the universe. They also thought that the Sun, Moon, and planets all circled around Earth, as did the stars fixed inside the celestial sphere. This was the classic Greek concept of the universe, elaborated by the last great Greek astronomer Ptolemy of Alexandria in about 150 A.D. This era was marked by the fabulous, fantastic tales of Greek mythology. These stories told of the epic adventures and wondrous deeds of gods and goddesses, heroes and heroines, fantastic creatures like centaurs, and monsters like Gorgons.

In Ancient Greece, astronomy and mythology intermingled, and the heavens became the immortal resting place for a host of mythological characters that were embodied in the constellations. We have inherited most of the Greek constellations, but we know them today not by their Greek names but by their Latin ones.

Between two rivers

The Greeks inherited many of the constellations from earlier times—predominantly from Mesopotamia. The Greeks gave the name Mesopotamia, meaning "between two rivers," to the region between the rivers Tigris and Euphrates in what is modern Syria and Iraq. This was where the Sumerians and Babylonians built up the first great civilizations in the Middle East, and where writing and the wheel were invented around 3000–3500 B.C. The first

The ancient Britons used Stonehenge to mark the passing of the seasons.

Ptolemy's idea of an Earth-centered universe, set out in an encyclopedic work known as Almagest *(*The Greatest*).*

Boundary stones were the equivalent of royal charters, and sought the protection of the gods for the ownership of the king's lands.

writing took the form of picture symbols, or pictographs. In pictographs on decorated pottery, in carvings and on seals, three figures were common— bulls, lions, and scorpions. These three figures were pictured in the sky in the earliest zodiacal constellations—the constellations the Sun passes through each year. They were the forerunners of Taurus, Leo, and Scorpius.

The heavenly triad

Later artworks showed other animals and gods, some clearly identified with heavenly bodies. Symbols of the Sun, Moon, and Venus—the three brightest heavenly bodies—became a recurrent theme. This triad was handed down to the Babylonians, where Venus was Ishtar, the "queen of heaven and whore of Babylon." Fine representations of these and the zodiacal constellations have survived on boundary stones of Babylonian times, dating from about 1300 B.C.

Ptolemy's Universe

Egyptian priest-astronomers were meticulous observers.

Arab astronomers developed many new instruments.

By Ptolemy's time, astronomers were recording the positions of the wandering stars—the planets—in the sky with some accuracy. And they realized that there was something odd about the motions of the planets. If they indeed circled around Earth as was believed, they should always travel in the same direction—toward the west, like the Sun. But often a planet could be seen to backtrack in the sky, moving eastward for some time before resuming its usual westward course.

So Ptolemy worked out a system of deferents and epicycles to try to account for this occasional backward motion. He said that a planet moved in a little circle (epicycle) around a point (deferent) that moved in a great circle around the Sun. This did not work very well either, and so over the years, further epicycles were introduced until the system became impossibly complex. And it still did not explain the motion of the planets satisfactorily.

Into and out of the Dark Ages

The problem of accounting for planetary motions seemed to be a fatal flaw in the Ptolemaic, Earth-centered view of the universe. Yet it was accepted, almost without question for nearly 1,400 years. In Europe, no one bothered much about astronomy anyway during this time. With the decline of Greco-Roman civilization by about the fifth century, Europe slipped into a period of general cultural stagnation when much of the knowledge of the ancient world was either lost or forgotten. This period is known as the Dark Ages.

Fortunately, astronomy continued to thrive elsewhere, particularly in Arabia. One of the triggers was the translation into Arabic of

The Sun was at the center of Copernicus' universe.

Ptolemy's seminal work the *Almagest*, circa 820 A.D. This inspired generations of Arab astronomers until 1428, when Ulugh Beigh established at Samarkand the finest observatory the world had ever seen.

It was also in the 1400s that the great rebirth of learning we call the Renaissance was getting underway in Europe. Philosophers and scholars began questioning and investigating age-old beliefs. In astronomy, a startling breakthrough came from an unlikely source—a cleric and physician named Nicolaus Copernicus.

Copernicus had a passion for astronomy. He was a keen observer himself and was familiar with other astronomers' observations. He came to realize that Ptolemy's concept of the universe was wrong. The odd movements of the planets could be explained simply if the Sun, and not Earth, was the center of the universe. The planets circled around the Sun, and so did Earth. Earth was merely another planet.

Nicolaus Copernicus (1473–1543) published his ideas about a solar system as he was dying.

De Revolutionibus

Galileo explains the sights seen through his telescope.

Copernicus was not the first person to put forward the idea of a solar system—a universe centered on the Sun. A Greek philosopher named Aristarchus, who lived in Samos, had done so around 200 B.C. But no one listened to him because his idea would have placed Earth in an inferior position in the universe.

Copernicus knew that his concept of a solar system would upset the Church, which at that time virtually dictated all opinions within society. Anything that went against Church beliefs was tantamount to heresy, and punishable by excommunication, torture, and even death. So it was not until Copernicus was on his deathbed in 1543 that he published his ideas in *De Revolutionibus Orbium Coelestium*.

The evidence mounts

An 18th century orrery —a mathematical model of the solar system.

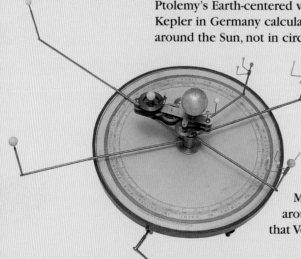

Opposed by the Church, the idea of a solar system took hold only slowly, and not before one ardent Copernican, Giordano Bruno, was burned at the stake in 1600. In 1609, two events took place that consigned Ptolemy's Earth-centered view to history. Johannes Kepler in Germany calculated that the planets travel around the Sun, not in circles but in elliptical (oval) orbits, which matched planetary observations exactly. Also in 1609, Galileo Galilei in Italy turned a telescope on the heavens and saw sights no one had ever seen before. He saw mountains on the Moon and moons circling around Jupiter. He also noticed that Venus showed phases, something that could happen only if the planet circled around the Sun. This provided convincing evidence of a solar system.

Background: Italian astronomer Galileo Galilei (1564–1642) pioneered telescopic observation of the heavens.

The expanding universe

As telescopes became more powerful, astronomers began to probe deeper into space. In 1781, William Herschel spotted a new planet, Uranus. It proved to be twice as far from Earth as Saturn, the most distant planet known to the ancients. At a stroke, the size of the solar system had doubled. It expanded further when Neptune was discovered in 1846, and Pluto as recently as 1930.

By the 1930s, the true nature of our universe had been pieced together. The solar system is a family of bodies centered on the Sun. The Sun is one of billions of stars that belong to a great galaxy, or star island, in space. There are many such star islands that group together into ever greater groups, or clusters, to form the universe. But the universe is not a static arrangement of galaxies scattered through space. All the galaxies are rushing away from one another. The whole universe seems to be expanding, as if from a primordial explosion eons ago. And astronomers believe that such an explosion did take place about 15 billion years ago. This Big Bang created the universe and set it expanding. Current evidence suggests that it is going to expand forever.

Englishman William Herschel (1738–1822), a musician turned astronomer, discovered Uranus in 1781.

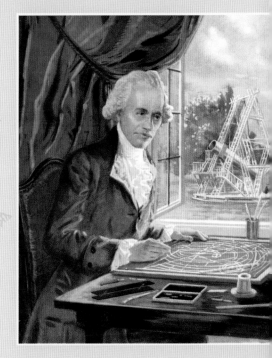

Chapter 2 · PATTERNS IN THE SKY

That the universe is expanding, there is no doubt. Neither is there any doubt that the stars themselves are shooting through space at incredible speeds and in all directions. But they are so far away that they hardly seem to move at all across the heavenly sphere. They appear fixed in their constellations. That is why Greek astronomers of 2,000 years ago would recognize the constellations that we see today as familiar friends. Only over periods measured in tens and hundreds of thousands of years do the stars change position enough to present new patterns to the eye.

According to Ptolemy, the Greeks recognized forty-eight constellations. They named them after figures they thought the patterns of bright stars formed in the sky. In a few cases, we can see what they saw so long ago. One constellation does look rather like a flying swan, another like a lion, and another like a scorpion. We know these and the other constellations by the Latin versions of the names the Greeks gave them. So the

Ursa Major, the Great Bear, whose brightest stars form the Big Dipper.

Ancient prints from 1515 showing northern (below) and southern (right) constellations.

swan is Cygnus, the lion is Leo, and the scorpion is Scorpius. In most instances, however, immense imagination is needed to associate a pattern of stars in the sky with a specific constellation name.

On with the new

Forty new constellations have been added since Ptolemy's time, making a total of eighty-eight in all. It might be expected that Arab astronomers would have introduced some since they, essentially, kept astronomy going through the Dark Ages. But they did not. Their main legacy has been in the names of many of the bright stars in the constellations, such as Betelgeuse in Orion, Algol in Perseus, Aldebaran in Taurus, and the beautiful Zubenalgenubi in Libra.

Three astronomers were mainly responsible for the additional constellations—in Germany, Johann Bayer in 1603 and Johann Hevelius in 1690; and in France, Nicolas Lacaille in 1752. It is to Bayer that we are also indebted for our system of identifying stars in a constellation by a letter of the Greek alphabet. They are graded in approximate order of brightness from alpha (brightest) onward.

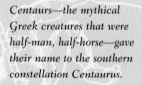

Centaurs—the mythical Greek creatures that were half-man, half-horse—gave their name to the southern constellation Centaurus.

The Hubble Space Telescope is forever probing the secrets of the constellations.

The Celestial Sphere

Long-exposure photographs of the night sky show trails made by the stars as they wheel overhead.

All the stars of the celestial sphere belong to our own star island in space: our galaxy.

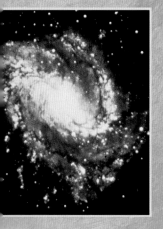

As we have noted, it seems as if the stars—the constellations—are stuck on the inside of a great celestial sphere surrounding Earth. And this great sphere seems to spin around Earth, making the stars travel across the night sky from east to west.

But just as the celestial sphere is an illusion, so is its motion. The heavens are not spinning, it is Earth that is in motion. Earth spins around in space once a day—this is in fact the way we define the time period we know as the day. The spinning of Earth is what makes the Sun appear to arc across the sky by day, and the stars to wheel across the sky at night. And Earth spins around in the opposite direction from the Sun and the stars—from west to east.

The celestial sphere appears to spin around on an axis (imaginary line) through the north and south celestial poles. These are points on the celestial sphere directly above Earth's North and South Poles. The celestial equator divides the sphere in two—into northern and southern celestial hemispheres. It is a projection on the celestial sphere of Earth's Equator, which divides Earth into the Northern and Southern Hemispheres. For convenience, we often divide the constellations in two according to the hemisphere in which they appear—northern or southern (see pages 18–19).

Pinpointing the stars

The concept of the celestial sphere may be ancient but it still has a role to play in modern astronomy. From an observer's point of view, it reflects exactly what we see in the heavens. In particular, it provides a simple means of pinpointing the stars in the sky. By using the geometry of a sphere, astronomers locate a star using a grid system analogous to the latitude and longitude

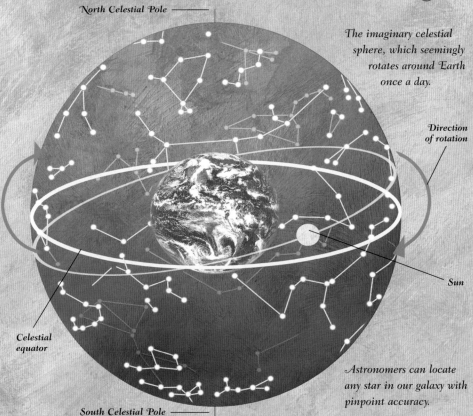

North Celestial Pole

The imaginary celestial sphere, which seemingly rotates around Earth once a day.

Direction of rotation

Sun

Celestial equator

South Celestial Pole

Astronomers can locate any star in our galaxy with pinpoint accuracy.

system geographers use for pinpointing a place on Earth. They pinpoint a star on the celestial sphere by its celestial latitude and longitude. The celestial latitude (called declination) is the angular distance of the star north or south of the celestial equator, just as latitude on Earth is the angular distance north or south of the Equator.

Celestial longitude is the distance around the celestial sphere from a fixed point (the First Point in Aries, see pages 30-31) just as longitude on Earth is the distance along the Equator from a fixed point (the Greenwich Meridian). Celestial longitude is known as right ascension (RA).

Northern Constellations

The constellations of the northern celestial hemisphere.

The darker area shown on the celestial sphere marks the Milky Way.

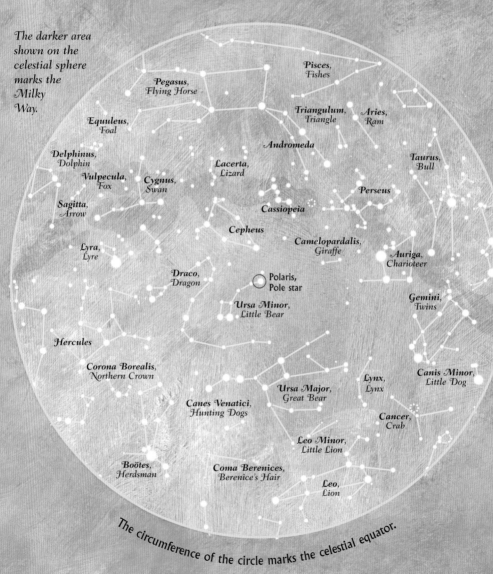

Pisces, Fishes

Pegasus, Flying Horse

Triangulum, Triangle

Aries, Ram

Equuleus, Foal

Andromeda

Taurus, Bull

Delphinus, Dolphin

Lacerta, Lizard

Vulpecula, Fox

Cygnus, Swan

Perseus

Sagitta, Arrow

Cassiopeia

Camelopardalis, Giraffe

Cepheus

Auriga, Charioteer

Lyra, Lyre

Draco, Dragon

Polaris, Pole star

Gemini, Twins

Ursa Minor, Little Bear

Hercules

Canis Minor, Little Dog

Corona Borealis, Northern Crown

Lynx, Lynx

Ursa Major, Great Bear

Canes Venatici, Hunting Dogs

Cancer, Crab

Leo Minor, Little Lion

Boötes, Herdsman

Coma Berenices, Berenice's Hair

Leo, Lion

The circumference of the circle marks the celestial equator.

Southern Constellations

The constellations of the southern celestial hemisphere.

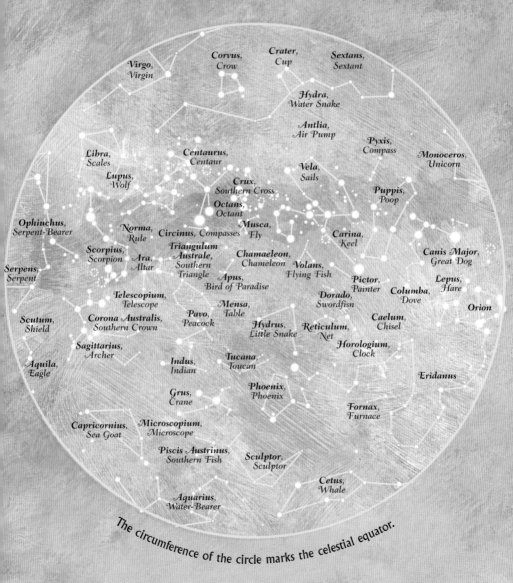

Virgo,
Virgin

Corvus,
Crow

Crater,
Cup

Sextans,
Sextant

Hydra,
Water Snake

Antlia,
Air Pump

Pyxis,
Compass

Monoceros,
Unicorn

Libra,
Scales

Centaurus,
Centaur

Vela,
Sails

Lupus,
Wolf

Crux,
Southern Cross

Puppis,
Poop

Octans,
Octant

Ophiuchus,
Serpent-Bearer

Norma,
Rule

Circinus, Compasses

Musca,
Fly

Carina,
Keel

Scorpius,
Scorpion

Triangulum
Australe,
Southern
Triangle

Chamaeleon,
Chameleon

Canis Major,
Great Dog

Serpens,
Serpent

Ara,
Altar

Volans,
Flying Fish

Lepus,
Hare

Apus,
Bird of Paradise

Pictor,
Painter

Columba,
Dove

Orion

Scutum,
Shield

Telescopium,
Telescope

Mensa,
Table

Dorado,
Swordfish

Caelum,
Chisel

Corona Australis,
Southern Crown

Pavo,
Peacock

Hydrus,
Little Snake

Reticulum,
Net

Aquila,
Eagle

Sagittarius,
Archer

Horologium,
Clock

Indus,
Indian

Tucana,
Toucan

Eridanus

Grus,
Crane

Phoenix,
Phoenix

Capricornius,
Sea Goat

Microscopium,
Microscope

Fornax,
Furnace

Piscis Austrinus,
Southern Fish

Sculptor,
Sculptor

Cetus,
Whale

Aquarius,
Water-Bearer

The circumference of the circle marks the celestial equator.

Astronomers begin stargazing when other people are thinking of going to bed.

Southern delights: the brilliant Alpha and Beta Centauri act as pointers to Crux, the Southern Cross.

What You Can See

Which constellations will you see when you go stargazing? That depends on many factors—where you are on Earth, the time of night, and also the time of year. Because Earth is spherical, people viewing from different latitudes will have different views of the celestial sphere—of the constellations.

If you stargaze from Toronto, Ontario, in Canada, high in the Northern Hemisphere, you will see different constellations than will fellow stargazers in Cape Town, South Africa, deep in the Southern Hemisphere. You will be able to spy the Big Dipper every night, but you will never get a glimpse of Crux, the Southern Cross (and vice versa, of course). This famous southern constellation is forever hidden by Earth's curvature.

Telling time

Which constellations you see also depends on the time of night that you are viewing. Because the Earth is spinning, the stars move across the sky all the time. Some constellations rise in the east as others disappear beneath the western horizon.

Different constellations also appear and disappear as the months go by. This is a result of Earth traveling in its orbit (path) around the Sun. Every day, Earth moves a little further along its orbit, and every night when you stargaze you look out at a slightly different part of the celestial sphere. This difference soon becomes noticable, and after three months—every season—the skies look quite different.

Orion is a dazzling constellation. The three bright stars in the middle outline Orion's belt. The bright patch beneath them is the finest nebula in the heavens.

Background: the constellation figure of Orion through the eyes of ancient astronomers.

Edwin Hubble (1889–1953) carried out most of his stargazing at Mt Wilson and Palomar Observatories, near Los Angeles.

This leads us to look at the spring, summer, fall, and winter constellations. For example, in the Northern Hemisphere, Orion is a spectacular winter constellation, while the Square of Pegasus warns that fall is on the way. In the Southern Hemisphere, the seasons are reversed, of course. So Orion is a feature of summer skies, while Pegasus promises spring.

As a rule

To sum up, an observer in the Northern Hemisphere, if he or she remained stargazing all year, would see all the constellations of the northern celestial hemisphere and some of the southern celestial hemisphere. In a similar way, an observer in the Southern Hemisphere would be able to see all the southern constellations and some of the northern ones. In theory, the best place for stargazing is on the Equator, where all the constellations would be visible at some time of the year.

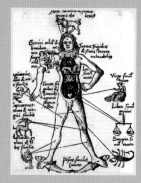

Ancient astronomers thought that the organs of the body were influenced by the stars.

Angels display the horoscope of Rudolf II, who became Holy Roman Emperor in 1576.

Constellations of the Zodiac

Earth travels through space in orbit around the Sun, making one complete journey every 365¼ days, or one year. But from our viewpoint on Earth, it appears that the Sun travels once around the heavens every year. We call the apparent path of the Sun around the celestial sphere the ecliptic. The Sun follows the same path every year against the background of stars. The Moon and the planets are also always found in the sky quite close to the ecliptic, within an imaginary band in the heavens called the zodiac.

The constellations the zodiac passes through are known as the constellations of the zodiac. "Zodiac" roughly translated means "circle of animals," referring to the fact that most of the constellations are named after animals, such as Leo (lion) and Scorpius (scorpion).

The believers

Because the Sun and planets were always found among the zodiac constellations, ancient astronomers considered that the zodiac must have special significance. And a belief grew up that somehow the positions of the Sun and planets within the constellations affected people's characters and the lives they led. This belief became known as astrology. It thrived for more than 2,000 years, from Babylonian

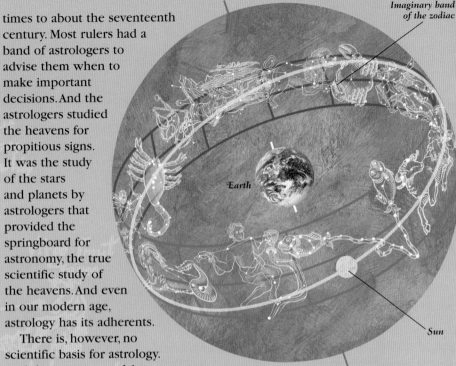

Imaginary band
of the zodiac

Earth

Sun

times to about the seventeenth century. Most rulers had a band of astrologers to advise them when to make important decisions. And the astrologers studied the heavens for propitious signs. It was the study of the stars and planets by astrologers that provided the springboard for astronomy, the true scientific study of the heavens. And even in our modern age, astrology has its adherents.

There is, however, no scientific basis for astrology. And there are powerful arguments against it. Astrologers base their belief on the position of the Sun among twelve constellations or "signs" of the zodiac—Aries, Taurus, Gemini, Cancer, Leo, Virgo, Libra, Scorpius, Sagittarius, Capricorn, Aquarius, and Pisces. But in fact there are thirteen constellations of the zodiac: the Sun passes through Ophiuchus on its way between Scorpius and Sagittarius.

Also, most modern day astrologers still assume that the Sun passes through each constellation at the same time as it did two millennia ago; but it doesn't. Because of precession—the slight gyration of the Earth's axis in space—the Sun passes through the constellations nearly a month earlier than it did in Roman times. This puts all the star signs out.

The imaginary band of the zodiac girdles the celestial sphere. The Sun, the Moon, and the planets are always found within it.

Chapter 3 · AROUND THE CONSTELLATIONS

Some constellations, such as Ursa Major (the Great Bear) in the Northern Hemisphere and Crux (the Southern Cross) in the Southern, are familiar even to the casual stargazer. Ursa Major is an ancient constellation, dating back to the beginnings of civilization in the Middle East, and has strong mythological associations. Crux is a more modern creation, which began to appear on star maps only in the sixteenth century.

This billowing cloud of gas in Cassiopeia is called the Bubble Nebula.

Dazzling star clusters in the Large Magellanic Cloud in Dorado.

Most of the thirty-three constellations included in this chapter date back to ancient times and are steeped in mythology. We will look at some of the myths behind the figures traced by these star patterns and mention a few observational highlights. The four brightest stars of each constellation are indicated by the Greek letters Alpha (α), Beta (β), Gamma (γ), and Delta (δ). If other stars are mentioned in the text, they are also identified by their Greek letter. A few stars are also named. Most of the names come from the Arabic, dating from the Middle Ages, when Arab astronomers led the world (see page 10). Familiar examples are Betelgeuse and Rigel, the two brightest stars in Orion, and Algol, the famous variable star in Perseus. There are Greek names, such as Sirius, the brightest star in the sky, and Latin ones such as Bellatrix, another bright star in Orion.

Mention is often made of magnitude, which

Features of the constellation spreads that follow:

Mythology relating to constellation

Astronomical feature of constellation

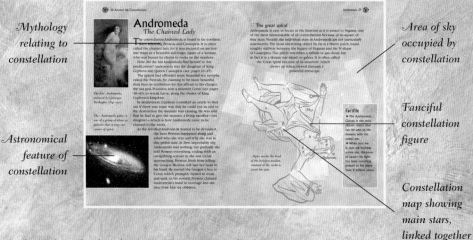

Area of sky occupied by constellation

Fanciful constellation figure

Constellation map showing main stars, linked together

describes the brightness of a star. In naked-eye astronomy, astronomers grade the stars in six magnitudes of brightness as we see them in the sky. (We call this their apparent magnitude). The brightest ones are first magnitude, the faintest ones sixth magnitude; the others have magnitudes in between. This magnitude scale is extended beyond six to cover the brightness of much fainter stars, visible only in telescopes, and into negative values to express the brightness of exceptionally brilliant objects.

Twelve of the constellations featured are the constellations of the zodiac, which provide the backcloth for the Sun's annual voyage around the celestial sphere. These are the constellations, or star signs, which astrologers revere and which, they say, shape our character and dictate our destiny.

The stories behind the constellations are invariably fascinating. Here, we find a wronged maiden turned into a bear by a jealous wife—there, a hero slaying a snake-headed ogre. All fantastic life is here in these stirring, often racy tales.

Meteor trails in Leo during the annual meteor shower called the Leonids.

Andromeda
The Chained Lady

The constellation Andromeda is found in far northern skies, adjoining Perseus and Cassiopeia. It is often called the chained lady, for it was pictured on ancient star maps as a beautiful and tragic figure of a woman, who was bound by chains to rocks on the seashore.

How did the fair Andromeda find herself in this predicament? Andromeda was the daughter of King Cepheus and Queen Cassiopeia (see pages 44-45).

The queen had offended some beautiful sea nymphs, called the Nereids, by claiming to be more beautiful than they. As retribution for this affront to his charges, the sea god, Poseidon sent a monster, Cetus (see pages 48-49), to wreak havoc along the shores of King Cepheus's kingdom.

In desperation, Cepheus consulted an oracle to find out if there was some way that he could put an end to the destruction the monster was causing. He was told that he had to give the monster a living sacrifice—his daughter—which is how Andromeda came to be chained to the rocks.

As the terrified Andromeda waited to be devoured, the hero Perseus happened along and asked who she was and why she was in this pitiful state. At first, improbably shy, Andromeda said nothing, but gradually she told Perseus everything, ending with an ear-splitting scream as she saw Cetus approaching. Perseus, fresh from killing the Gorgon Medusa, still had her head in his hand. He turned the Gorgon's face to Cetus, which promptly turned to stone and sank. As his reward, Perseus claimed Andromeda's hand in marriage and she later bore him six children.

The fair Andromeda, pictured by Guiseppe Barbaglia (1841–1910).

The Andromeda galaxy is one of a group of about 30 galaxies that occupy our corner of space.

⅝ The great spiral

Andromeda is easy to locate in the heavens as it is joined to Pegasus, one of the most unmistakable of all constellations because of its square of four stars. Visually, the individual stars in Andromeda are not particularly noteworthy. The most interesting object by far is a blurry patch, found roughly midway between the Square of Pegasus and the W-shape of Cassiopeia. This patch resembles a nebula or gas cloud, but in fact it is a distant star island, or galaxy. It is often called the Great Spiral because of its structure, which shows up when viewed through a powerful telescope.

Alpha marks the head of the helpless maiden, chained to the rocks to await her fate.

Factfile

★ The Andromeda Galaxy is the most distant object that can be seen in the heavens with the naked eye.

★ When you see it, you are looking across vast distances of space—its light has been travelling toward us for more than 2 million years.

Aquarius
The Water-Bearer

Feluccas on the Nile, the river that was the life-blood of the early Egyptians.

The tadpole-like "cometary knots" in the Helix nebula, that come from a dying star.

The entire region in which Aquarius lies has watery connections—the constellation is surrounded by the Fishes, the Sea Goat, the Southern Fish, and the Whale. The Babylonians called this region "the Sea", with all constellations being under the control of Aquarius. The Egyptians believed that Aquarius caused the annual flooding of the Nile River, and so it was a very important constellation to them. It is no coincidence that the hieroglyph for running water is now the astrological sign for Aquarius.

Traditional star maps depict Aquarius as a youth pouring water from a pitcher. Beneath his feet, the gushing water ends up in the mouth of Piscis Austrinus, the Southern Fish. The beautiful youth was probably Ganymede, son of the King of Tros, founder of the city of Troy. Zeus wanted Ganymede as his favorite and sent an eagle to carry him off to Mount Olympus. There he became the cupbearer of the gods, delighting all with his beauty.

Starry triangle

Aquarius is not an easy constellation to make out. It can probably best be located by reference to the Square of Pegasus to the north. Even its lead star Alpha (α) is disappointingly dim. There are a few highlights, however; one is M2, which makes a triangle with the stars Alpha and Beta (β).

Alpha and Beta mark
the shoulders of the
figure, who the Greeks
associated with the
handsome Ganymede.

It is one of the
brightest globular
clusters in the sky
and just on the limit
of naked-eye visibility. Binoculars and
small telescopes show it well.

〰 What the astrologers say

Aquarius is a constellation of the zodiac lying between
Pisces and Capricornus. Currently, the Sun passes
through Aquarius from February 16 to March 11.

In astrology, Aquarius is the eleventh sign of the
zodiac, covering the period January 20 to February
18. Astrologers believe that Aquarians are cool and
detached, even aloof; they tend to conceal emotions
and find it difficult to fall in love or relate to partners.

Factfile

✦ Larger telescopes
are needed to spot
the Helix nebula,
near the southern
end of the
constellation.

✦ It is one of the
nearest planetary
nebulae and the
Hubble Space
Telescope has
discovered some
incredible "cometary
knots" in it.

Aries
The Ram

Factfile

★ The two main stars of Aries are of the second magnitude, but it is the fainter Gamma that is the more noteworthy.

★ In small telescopes, Gamma shows up as a lovely double, with the pair of stars of equal blue-white brilliance.

Movement of the First Point of Aries 1700-2300 A.D. It currently lies in Pisces.

A ries is a small constellation, and is not particularly easy to find. It is best located by following a line from the unmistakable cluster of the Pleiades to the Square of Pegasus. It is located about halfway between the two.

In mythology, the ram was the source of the fabulous Golden Fleece that Jason and the Argonauts went to steal (see pages 42–43). The ram was a magical creature that could speak, think, and fly through the air. Hermes gave the ram to the two children of King Athamas, Helle and her brother Phrixus, who fled on it to escape from their hated stepmother. Helle unfortunately fell off into a strait that became known as the Hellespont. Phrixus safely reached Colchis on the Black Sea and sacrificed the ram to show his gratitude for being saved. He gave the Golden Fleece to Aeetes, king of Colchis, who set the ferocious dragon Draco, that never slept, to guard it.

♈ What the astrologers say

Aries is a constellation of the zodiac, lying between Pisces and Taurus. Currently, the Sun passes through Aries between April 18 and May 21.

In astrology, Aries is the first sign of the zodiac, covering the period from March 21 to April 19, for that is what it did in classical times. March 21 is around the time of the vernal, or spring, equinox, when the Sun crosses the celestial equator travelling north. On this date, the path of the Sun through the heavens (the ecliptic) and the celestial

Pisces

0^h

Ecliptic

1700
1800
1900
2000
2100
2200
2300

Celestial Equator

Aquarius

The three brightest stars —Alpha, Beta, and Gamma—outline the head of the ram with the golden fleece.

α

ε

δ

θ

β

γ

ι

ξ

equator intersect. This point is known as the First Point of Aries. However, because of precession (a wobbling of the Earth's axis), the Sun is no longer in Aries at this time, but in Pisces (see pages 78–79).

Astrologers tell us that Arians are energetic, impulsive, and extroverted. They are believed to be courageous and adventurous, and natural-born leaders. Warm and passionate .in relationships,they frequently find the chase more exciting than the surrender.

Jason and the Argonauts encountered many dangers on their quest for the Golden Fleece.

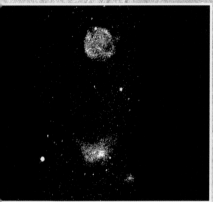

Near Capella, x-rays reveal two supernova remnants (SNRs), the remains of the exploding stars.

Factfile

★ Auriga's lead star, Capella, is nearly 100 times brighter than the Sun, a yellow dwarf less than one-tenth the size of Capella.

★ Capella is a spectroscopic double—the stars are so close together that they can be detected only in a spectroscope.

Auriga
The Charioteer

This familiar northern constellation depicts the driver of a horse-drawn chariot, and originally included the chariot as well. Auriga is usually identified with Erichthonius, who became king of Athens. His father was the Greek god of fire, Hephaestus (Vulcan in Roman mythology), but he had no mother.

According to the myth, this came about because Hephæstus had lusted after the conspicuously virginal goddess Athene, and one day attempted to seduce her. She fought him off and escaped, while he scattered his seed on the earth. Shortly afterward, the earth gave birth to a boy-child, Erichthonius. Then, by chance, Athene found him and brought him up, teaching him the skills of horsemanship. Erichthonius then developed the four-horse chariot. Later he became king of Athens and established the cult of Athene worship there in gratitude.

Auriga is usually depicted with a goat draped over his left shoulder. The brightest star in the constellation, Capella marks its position—the name means she-goat. The goat is generally identified with Amalthea, the goat that suckled Zeus as a baby. The goat's kids are also present, held by Auriga's left arm. They are marked by a triangle of fainter stars and are known as the *Kids*, or the *Heidi*.

☝ The Goat Star

Auriga is not difficult to spot because of its lead star Capella. It forms a compact triangle of constellations with Gemini and Taurus. Capella is often referred to as the Goat Star. It is the sixth brightest star in the sky and it emanates a yellowish light similar to that given off by

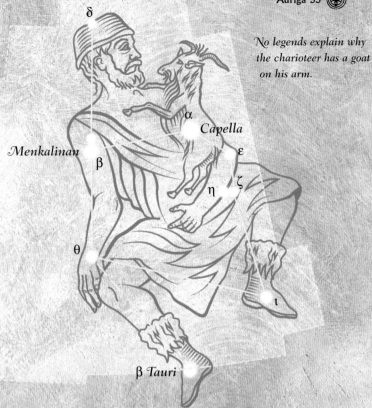

δ

No legends explain why the charioteer has a goat on his arm.

α

Capella

Menkalinan

ε

β

η ζ

θ

ι

β *Tauri*

our own Sun. But there the resemblance ends. Two of the three stars that mark the Kids, Epsilon (ε), and Zeta (ζ) are also binary, or double-star systems. They are eclipsing binaries, whose brightness periodically dims when one of the stars eclipses the other. Since Auriga straddles the Milky Way, it is a good subject for sweeping with binoculars. Several bright open clusters at the edge of the Milky Way, near the star Theta (ϑ) are readily visible.

Erichthonius' four-horse chariot won him the admiration of the gods.

Boötes
The Herdsman

This prominent northern constellation has an unmistakable shape of a traditional kite. It is easily located by following the curve of the handle of the Big Dipper (the Plow), the prominent part of the Great Bear (Ursa Major). Interestingly, bears also feature in the story of this constellation. It was sometimes known as Arctophylax, meaning the Bear Keeper. And Boötes was sometimes referred to as the Bear Driver, rather than the Herdsman, chasing the Great Bear and the Little Bear (Ursa Minor) across the sky. Boötes holds the leads of the hunting dogs, represented by the constellation Canes Venatici, that pursue the bears across the sky.

On this 17th-century celestial globe, Boötes holds the hunting dogs (Canes Venatici) on leads.

Boötes is usually thought to represent Arcas, who was an offspring of Zeus and the nymph Callisto. Callisto had earlier taken a vow of chastity and became a companion of Artemis. However, cunning Zeus appeared to her in the guise of Artemis, and when she discovered his real identity, it was too late. So she gave birth to Arcas. Artemis was furious with Callisto, and Zeus changed her into a bear to help her escape Artemis's rage.

One day, when he was a man, Arcas was out hunting and came across Callisto in the guise of a bear. She recognized him, but he did not know his mother. He gave chase and she sheltered in a sacred place in which discovery would mean death. To protect her and Arcas, Zeus placed them in the heavens. There are several variations of this story, and indeed other stories about the origin of the constellation.

⛅ The bear's tail

Arcturus translated means bear's tail, and it is found at the tail end of the constellation's kite shape. Of first magnitude, it is a red giant, nearing the end of its life, and to the eye has a definite reddish tinge.

There is little else of naked-eye interest in this constellation, but it has some fine double stars visible through small telescopes. The best must be Epsilon (ε), seen as a beautiful yellowish-orange and bluish-green pair. Xi (ξ) and Mu (μ) are also colorful pairs.

Factfile

★ Boötes' leading star, Arcturus, is the brightest in the Northern Hemisphere and the fourth brightest in the whole sky.

★ About twenty-five times the diameter of the Sun, Arcturus is also 100 times brighter.

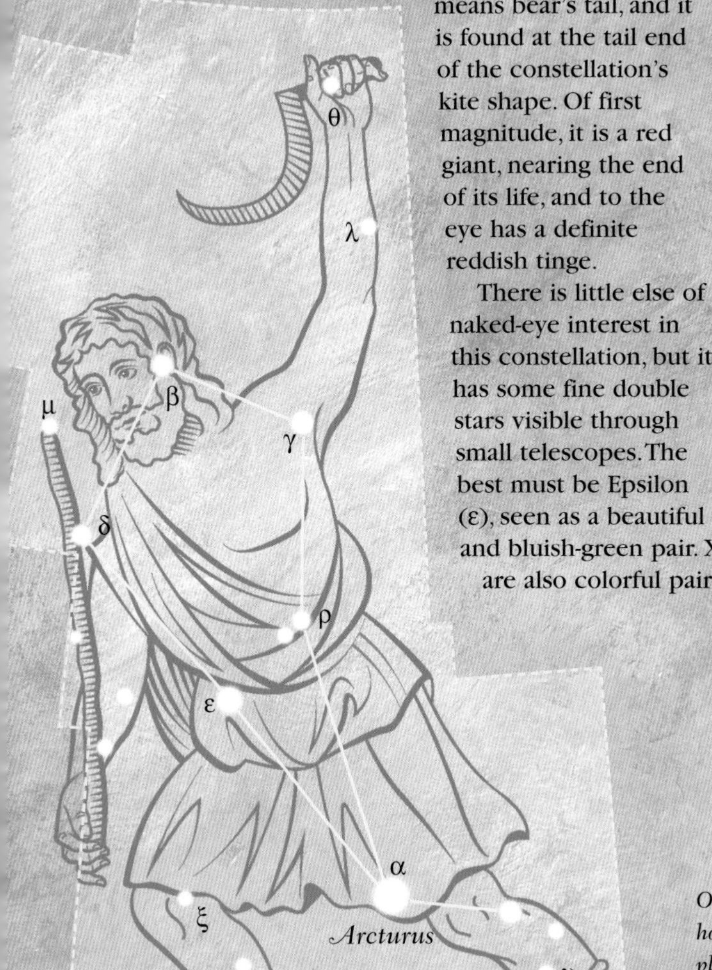

θ

λ

μ β

γ

δ

ρ

ε

α

ξ *Arcturus*

υ

One ancient story tells how Boötes gained his place in the heavens for inventing the ox-drawn plow.

Præsepe (M44), the fine open star cluster at the heart of the Crab.

Cancer
The Crab

Cancer is one of the faintest of the constellations but can be found quite easily east of a line connecting the bright stars in Gemini—Castor and Pollux—with Procyon in Canis Minor to the south.

In mythology, Cancer represents a crab that became involved in battling with Hercules while he was fighting the dreaded hydra. Hercules was the son of Zeus and the nymph Alcmene, the product of one of the king of the gods' many brief illicit affairs. Zeus's wife, Hera, thereafter hated Hercules and did her best to destroy him, first making him kill his wife and children in a fit of madness, and then getting him involved in virtually-impossible tasks, or labors, which would surely kill him.

Having defeated the Nemean lion on his first labor, Hercules took on the multiheaded hydra and vanquished this monster as well. In a rage, Hera sent a crab to attack Hercules, but to little avail. He crushed the crab underfoot, whereupon Hera placed it in the heavens.

Bees around the hive

The most interesting feature of this faint constellation is centered on the middle star, Delta (δ). If you look at Delta on a really dark night, you can see a swarm of stars close by. This swarm is named Præsepe, and is called the Beehive because the stars look rather like bees swarming around the hive.

What the astrologers say

Cancer is one of the constellations of the zodiac, lying between Gemini and Leo. Currently, the Sun passes through Cancer between July 20 and August 10. In astrology, Cancer is the fourth sign of the zodiac,

Factfile

✦ Near Cancer's middle star, Delta, lies an open cluster of stars called Præsepe. It looks better viewed with binoculars and best using a small telescope.

✦ In all, Præsepe probably contains at least 300 individual stars.

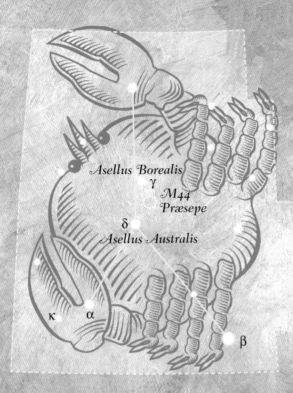

Asellus Borealis
γ
M44
Præsepe
δ
Asellus Australis

κ   α

β

The Italian astronomer Galileo first resolved the Præsepe cluster into individual stars.

Hercules gets to grips with the hydra. After battling this monster, he had no problem in killing the crab Hera sent to bite him.

covering the period from June 22 to July 22, for that is what it did in classical times. June 22 is around the time of the summer solstice, when the Sun reaches its furthest point north of the Equator. This extremity defines the northernmost point of the Tropics, and is called the Tropic of Cancer. However, because of precession, the Sun is now in Gemini at the summer solstice, so in theory it should now be renamed the Tropic of Gemini.

Astrologers say that Cancerians tend to be emotional people, quiet and even sullen at times. But they are also caring, kindly people; home-loving and dependable, they make good partners, but can become vindictive if they are crossed in love.

Factfile

★ Sirius, Canis Major's lead star, appears so bright to us because it lies relatively close—a little under nine light-years away.

★ Sirius is only about twenty-five times brighter than our Sun, while Rigel, lying more than 100 times further away in Orion, is some 60,000 times brighter.

Canis Major
The Great Dog

Canis Major represents one of the dogs that accompanied the hunter Orion. In the heavens it spends its time chasing the hare (the constellation Lepus) at Orion's feet. Orion's other dog, Canis Minor (the Little Dog), is located across the Milky Way from Orion and Canis Major. In mythology, this constellation is associated with the dog Lælaps, which could outrun anything it chased. Eventually, though, it went in pursuit of a fox that ran so fast that nothing could catch it, and the two seemed destined to race each other forever. But Zeus stepped in and turned them both to stone, placing Lælaps in the heavens as Canis Major.

The brightest star of all

The constellation is dominated by its lead star Sirius, which is the brightest star in the heavens. Unsurprisingly, it is popularly known as the Dog Star, and has possessed this name since classical times. The name Sirius means scorching. The Greeks thought that it was responsible for the heat of summer, since in July and August it rose in the dawn sky just before the Sun. These hot, sultry days were named the Dog Days.

The early Egyptians did not recognize any dog constellation, but they regarded Sirius as vitally important. They called it Sothis, or the Nile Star. They even worshiped it because it happened to appear in the dawn sky just before the Nile flooded each year, which they depended on to irrigate and nourish their land for agriculture.

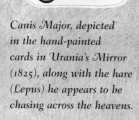

Canis Major, depicted in the hand-painted cards in Urania's Mirror (1825), along with the hare (Lepus) he appears to be chasing across the heavens.

In ancient Egypt, this constellation was depicted as a cow lying down, with Sothis (Sirius) between its horns.

Sirius has a magnitude of about -1.5 (negative magnitudes indicate exceptionally bright stars). It is about twice as bright as the next brightest star, Canopus. Strangely, both these stars lie relatively close together in the heavens, though this is just a coincidence.

Sirius is not a single star, but a binary. The other star is a tiny white dwarf. Indeed it was the first star of this type to be discovered, in 1862. It is named the Companion of Sirius, with the popular nickname of *the Pup*.

At x-ray wavelengths, the tiny white dwarf companion of Sirius appears brighter than Sirius itself. It is intensely hot, with a temperature of about 25,000°C.

Capricornus
The Sea Goat

The Greek god Pan, teaching one of his followers to play the pipes he invented.

Capricornus is a relatively faint constellation, but it is one of the most ancient and is always associated with water and the oceans. In Babylonian times, it was known as the Goat-Fish and was considered to rule the surrounding heavens, from which the great rivers Tigris and Euphrates flowed.

The Greeks also saw Capricornus as a strange creature that had the head and forelimbs of a goat but the tail of a fish. They identified it with the pipe-playing god Pan, who was also a strange hybrid creature with goat's legs and horns. Pan was the great god of the forests and meadows, through which he forever wandered, playing and dancing with the nymphs. One day he was pursuing the nymph Syrinx with amorous intent, but her sisters turned her into reeds as he was about to pounce. As he sighed, his breath blew over the reeds, which gave off musical sounds. He cut off a number of them of different length and bound them together to form the pipes of Pan, which are also called the syrinx.

♑ What the astrologers say

Astrologers always call this constellation Capricorn. It is the tenth sign of the zodiac, covering the period from December 22 to January 19, as in classical times. Currently, the Sun passes through Capricornus between January 19 and February 16.

December 22 is near the time of the winter solstice, when the Sun reaches its furthest point south of the

Equator. This extremity defines the southernmost point of the Tropics, and is called the Tropic of Capricorn. However, because of precession, the Sun is now in Sagittarius at the winter solstice, so we should now really refer to the Tropic of Sagittarius.

According to astrologers, Capricorns tend to be hard-working and reliable. They may have a hard exterior, but tend to be sensitive underneath. They crave respect and recognition. They are loyal but may sometimes seem rather cool partners, reluctant to show their true feelings.

The goat-fish, pictured on a sarcophagus at Thebes.

Factfile

✦ Capricornus is an inconspicuous constellation with only a few third-magnitude stars.

✦ It is the smallest constellation of the zodiac, lying in a "watery" region of the heavens between Sagittarius and Aquarius.

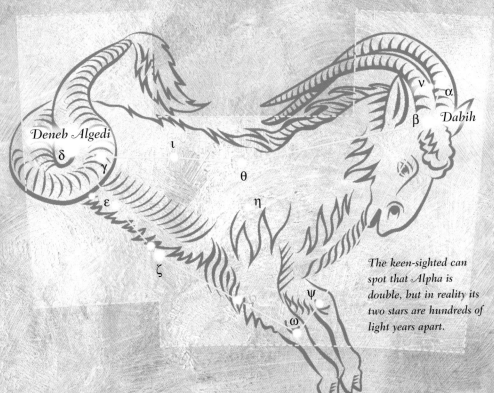

Deneb Algedi

δ

γ

ι

ε

ζ

θ

η

ν

α

β *Dabih*

ψ

ω

The keen-sighted can spot that Alpha is double, but in reality its two stars are hundreds of light years apart.

Carina
The Keel

T his is one of the stunning far southern
constellations beyond the gaze of most
Northern Hemisphere astronomers.

The ancients did not recognize Carina
as a separate constellation, but as part of
the much larger constellation Argo Navis,
the Ship of the Argonauts. Carina represents
the keel of this craft, while Vela represents
the sails and Puppis the poop (or stern).

*Argo Navis depicted in
an 18th century star map,
showing the ship slipping
through the Symplegades,
or Clashing Rocks.*

The French astronomer Nicolas Lacaille split the
constellation into three parts in 1763.

The Argonauts feature in one of the most famous
legendary tales of Ancient Greece—Jason's quest for the
Golden Fleece. The Golden Fleece was found at Colchis,
in a cave guarded by the dragon Draco. Jason took with
him fifty heroes, including Castor and Pollux (the twins
in the constellation Gemini, see page 60–61) and the
musician Orpheus.

On their journey to and from Colchis, Jason and his
crew had many adventures. For example, they had to
pass through the Clashing Rocks, which opened and
shut like sliding doors and would crush anything
caught between them. Jason got through safely by
sending a dove ahead to make the Rocks shut, then

*Swirling hot gas and cold,
dark clouds are revealed
in this image of the
Keyhole Nebula, or
Eta Carinae.*

rowed for dear life when they briefly
opened up again. At Colchis, Jason fell
in love with Medea, daughter of King
Aeetes, who possessed the Golden
Fleece. She was a sorceress and helped
the Argonauts steal it and returned
with them to Greece. Jason left the
ship Argo in a grove sacred to
Poseidon at Corinth.

Sailing the Milky Way

Carina and the rest of Argo Navis are set in one of the most dazzling regions of the Milky Way. It adjoins Crux, the Southern Cross, and Centaurus, the Centaur. In the region of the Milky Way, halfway between the False Cross and the Southern Cross, is one of the most brilliant nebulae in the heavens, the Eta Carinae Nebula NGC3372. In the 1800s, the star Eta (η) Carinae itself flared up to become brighter than Canopus. It is one of the largest of the known stars and is highly unstable, constantly puffing out vast clouds of gas and dust.

Factfile

✦ Carina's brightest star, Canopus, lies away from the Milky Way—only Sirius shines more brightly.

✦ In the central part of Carina, two stars form a cross with two stars in the adjacent Vela. Stargazers can confuse this grouping with the Southern Cross, and so it is called the False Cross.

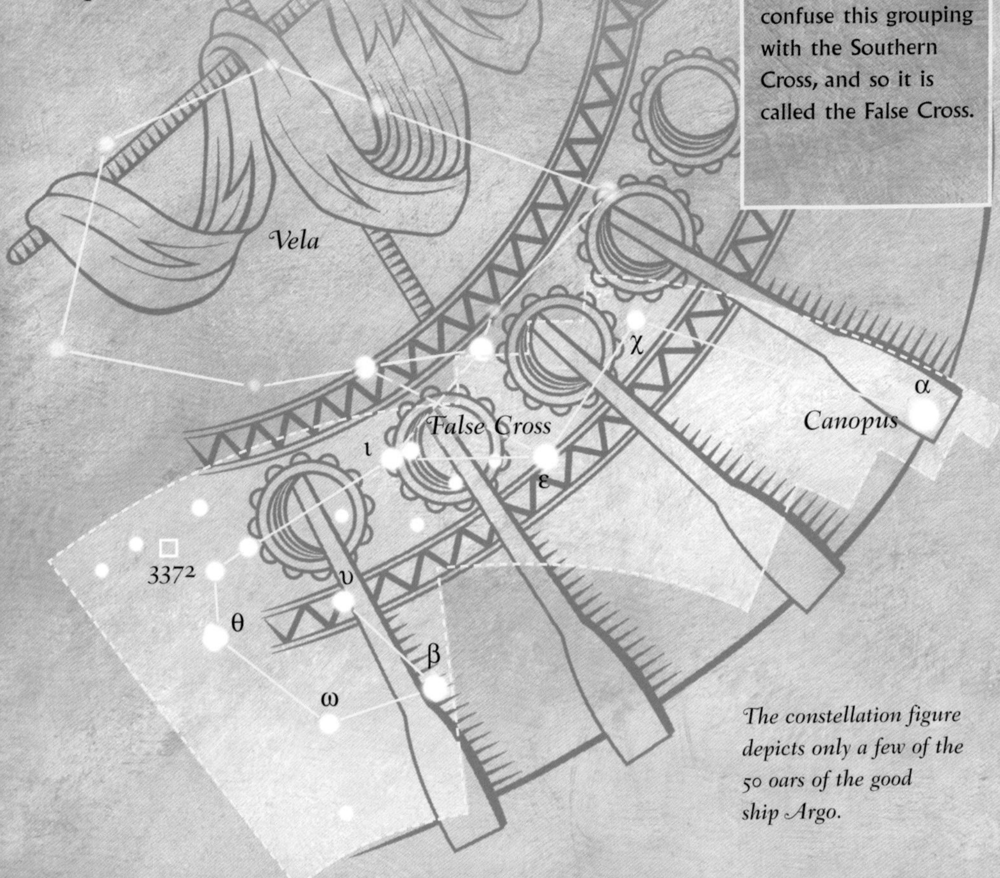

Vela

χ

α

Canopus

False Cross

ι

ε

3372

υ

θ

β

ω

The constellation figure depicts only a few of the 50 oars of the good ship Argo.

Cetus
The Sea Monster or Whale

This constellation is the fourth largest in the
heavens and it is an ancient one—Ptolemy listed
it as having twenty-two stars nearly two millenia ago.
Cetus is part of the ocean-related group that spans a
vast expanse of the heavens, which also includes
nearby Pisces (the Fishes, see pages 78–79), Aquarius
(the Water-Bearer, see pages 28–29), and Piscis
Austrinus (the Southern Fish).

Cetus has always been depicted as a bizarre-looking
monster of tremendous size and horrific appearance.
Its enormous head has gaping jaws studded with
vicious teeth, and its stubby forelimbs end with sharp
claws. Its long body is covered with scales, and its tail
is coiled like that of a sea serpent.
Although Cetus is often identified with
the Whale, there is nothing whale-like
about this creature, apart from its huge

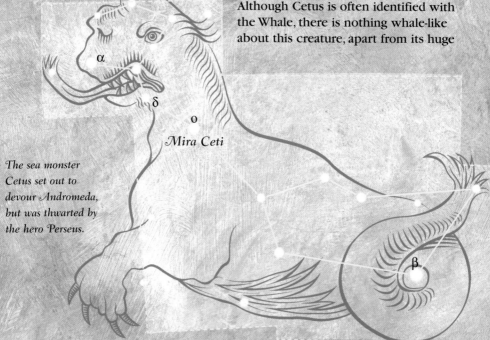

α

δ

o

Mira Ceti

β

*The sea monster
Cetus set out to
devour Andromeda,
but was thwarted by
the hero Perseus.*

size. Ancient peoples would have been familiar with whale skeletons even if they never saw a whale in the flesh, so they undoubtedly clothed such skeletons in their imagination and came up with the monster.

This monster "swims" in the heavens not so far from Andromeda, representing the chained maiden who in mythology was presented as a sacrifice for it. The sacrifice had been made because her boastful mother had offended the gods (see page 26–27). Before Cetus could get at Andromeda, he was killed by the hero Perseus (see page 44). One story says it was put to the sword by Perseus, another that it was turned to stone when Perseus exposed it to the eyes of Medusa, whose head he was carrying.

Perseus fights the monster Cetus, painted in 1515 by the Florentine artist Piero di Cosimo.

Mira the Wondrous

Cetus is a large and sprawling constellation, but a faint one. Its two brightest stars Alpha (α) and Beta (β) are about third magnitude. They are set respectively in the head and tail of the monster.

About a third of the way along the line from Alpha to Beta is the most interesting star for observers. It is designated Omicron (o) in the constellation, but astronomers also call it Mira, meaning the Wonderful One. Mira Ceti is wonderful because it varies in brightness markedly over time. At its brightest it is third magnitude, like Alpha and Beta, and is easily visible to the naked eye. But then it gradually fades to about magnitude 10, which renders it invisible to the naked eye—even with ordinary binoculars. Later it gradually regains its former brightness.

Mira is the prototype of the long-period variable stars, that are known as Mira variables. It is a huge red giant, which varies in brightness over a period of about eleven months.

Cygnus
The Swan

Cygnus is one of the few constellations that looks passably like the figure it is supposed to represent. With only a little imagination one can flesh out its pattern of bright stars into a swan, with wings extended in flight and long neck outstretched.

To the Greeks, Cygnus represented the king of the gods, Zeus, in disguise. Among the many women he lusted after was Leda, the queen of Sparta. He turned himself into a swan and slept with her on the riverbank. Later that night, Leda also slept with her husband, the king. Leda became pregnant, but instead of giving birth to babies, she produced two enormous eggs.

From one egg hatched Castor and Pollux (the twins in Gemini, see pages 60-61); from the other Clytemnestra and her sister Helen. Castor and Clytemnestra were fathered by the Spartan king, Pollux and Helen by Zeus.

Helen grew to be the most beautiful woman in the world, the famed Helen of Troy, with "the face that launched a thousand ships and sank the topless towers of Illium." This refers to the fact that she was the cause of the Trojan Wars between Sparta and Troy.

The Swan's tail

The shapely North American Nebula is a huge cloud of glowing hydrogen gas.

Astronomically, Cygnus is a delight which looks beautiful to the naked eye, flying along the Milky Way. Its brightest star, Deneb, in the Swan's tail, is first magnitude. It lies much further away than most other bright stars. But we still see it shining brilliantly because it is exceptionally luminous, shining with the power of more than 60,000

Deneb

α

ξ

δ

γ

ε

Albireo
β

*It is not surprising
that an alternative
name for this
constellation is the
Northern Cross.*

suns. Deneb is one of the three bright stars that make
up the celebrated Summer Triangle, along with Vega in
Lyra and Altair in Aquila.

The Milky Way is as always worth sweeping in
binoculars. The region between Deneb and the fainter Xi
(ξ) reveals a large nebula. But telescopes are needed to
bring out its uncannily life-like shape of North America,
and it is indeed called the North American Nebula.

*The god Zeus wooed Leda
in the guise of a swan.*

Spidery-looking wisps of glowing gas form the Tarantula Nebula in the Large Magellanic Cloud (LMC).

Dorado
The Swordfish

Canopus

Dorado is one of the more modern constellations, first depicted on star maps in the late sixteenth century. It lies in a fairly undistinguished part of the far southern heavens, not far from the south celestial pole.

The constellations surrounding Dorado—Mensa (the Table), Pictor (the Painter), Caelum (the Chisel), Reticulum (the Net), and Hydrus (the Little Snake)—are also recent in origin and in the main have little to interest the casual observer. There is no classical mythology associated with these modern constellations, which were created out of what are called amorphae, or the "left-over" stars.

Star island in space

The line of fairly faint stars that marks out Dorado can best be found by locating the beacon star Canopus and moving in the direction of the first magnitude Achernar in Eridanus. Slightly to the south of this line you will see Dorado's main claim to fame; a blurry patch, easily visible to the naked eye.

This blurry patch looks like a rather detached section of the Milky Way, but it isn't. Binoculars cannot resolve it into stars like they can the Milky Way. But small telescopes will begin to spot in it bright stars, and large telescopes will pick up masses of fainter stars. It seems to be a star system much farther away than the Milky Way. And so it is. It is another star island in space—another galaxy. In fact, it is the next nearest galaxy to our own.

Cloud-like to the naked eye, it is known as the Large Magellanic Cloud (LMC). Very

The LMC also boasts this double cluster of dazzling stars.

Dorado's gem, the Large Magellanic Cloud, lies about 170,000 light years away from us.

γ

α

β

δ

Large Magellanic Cloud

Factfile

✶ The Large Magellanic Cloud in Dorado contains much the same mix of ordinary stars, variables, clusters, and nebulae as in our galaxy.

✶ One of these nebulae can be seen with the naked eye. It is called Tarantula Nebula, after its resemblance to the notorious spider.

This stellar nursery of newborn stars is found within the Small Magellanic Cloud.

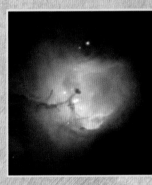

Small Magellanic Cloud

much smaller than our own galaxy and irregularly shaped, the LMC is named after the famed Portuguese navigator Ferdinand Magellan, who would have been one of the first Europeans to see it, it is our closest galactic neighbor. He commanded the first expedition to sail around the world, which set out in 1519. However, Magellan never completed this first circumnavigation as he was killed in the Philippines two years later.

And why is it known as the "Large" Magellanic Cloud? Well, following a line from Canopus through the LMC, you come to a smaller fuzzy patch. This is the Small Magellanic Cloud, another slightly more distant galactic neighbor.

Draco
The Dragon

Hercules dallying with the Hesperides in their wondrous garden.

The pyramids were built when Thuban (Alpha Draconis) was the pole star.

Draco winds itself, serpent-like, nearly halfway around the celestial north pole, marked by Polaris, the Pole Star. It is so far north that from most of the Northern Hemisphere, it is circumpolar, which means that it never sets and is always visible over the horizon.

In mythology, the dragon that never sets became the dragon that never sleeps. Known as Ladon, this monster had 100 heads and was an eternally vigilant guardian of the sacred golden apples in the exquisite Garden of the Hesperides, located in the most remote western limits of the world. The Hesperides were the daughters of Atlas and Hesperus, the evening star (Venus). They had originally been entrusted with the task of guarding the apples, but could not resist picking and eating them.

As the eleventh of his Labors, the most heroic of the Greek heroes, Hercules, set out to steal the golden apples for his cousin, Eurystheus, King of Argos. In seeking the garden, he had many brushes with death having to overcome many attackers, including a lion and a terrifying eagle. Eventually he arrived at the Garden of the Hesperides. There he slew Ladon and made off with the apples. One version of the story says that Atlas, who carried the world on his shoulders, helped him. Hercules persuaded Atlas to pick the apples while he temporarily carried the world on his shoulders.

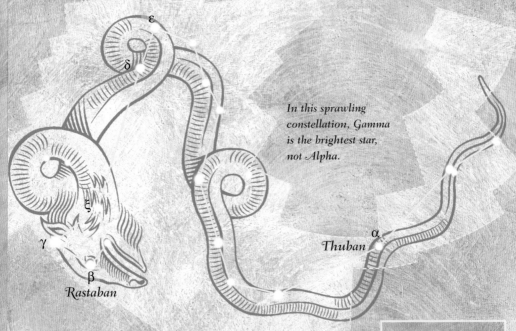

Polaris

ε

δ

In this sprawling
constellation, Gamma
is the brightest star,
not Alpha.

ξ

α

γ

Thuban

β

Rastaban

Thuban and Pole Stars

Draco winds its way around Polaris. But Polaris has
not always been the Pole Star. As the Earth spins
around in space, it wobbles, rather like a spinning top.
And this wobbling motion, known by astronomers as
precession, makes the Earth's axis point in different
directions in space at different times.

At present, the axis is pointing to Polaris, so this is
now the Pole Star. But some 4,000 years ago in Ancient
Egypt, the axis pointed at the star Alpha (α) in Draco,
which is also called Thuban (meaning serpent's head).
Thuban will again become the Pole Star in about
22,000 years, for it takes the Earth some 26,000 years
to go though each cycle of precession.

Factfile

★ Draco is a
sprawling serpent-like
constellation—the
eighth largest in
the heavens.

★ The Great
Pyramid of Giza,
built around the
time when Thuban
was the Pole Star, has
a passage lined up
with it.

Hydra
The Water Snake

O ut of all 88 constellations, Hydra is the largest. It winds its way, serpent-like, more than a quarter of the way around the celestial sphere. It runs roughly parallel to the celestial equator, south of the three constellations of the zodiac, Cancer, Leo, and Virgo.

In mythology, the hydra was a nine-headed serpent, born of the monster Typhon. The middle head was immortal. It inhabited the marshes near Lerna, in the Peloponnese, ravaging and terrorizing the surrounding region. It had deadly poisonous breath so those who felt it died an agonizing death.

The hero Hercules was set the task of killing the hydra as the second of his twelve labors. Accompanied by his chariot-driver Iolaus, Hercules arrived at Lerna and found the monster at the spring of Amymone. By firing flaming arrows into the marshes, he forced the hydra out into the open. He sprang at it, smashing his club into the serpent's heads. But every time he struck off one of the heads, two new heads grew back in its place.

Iolaus then set the surrounding forest on fire and with flaming brands, burned off the stump of each severed head, until only the immortal head remained. Hercules cut this off

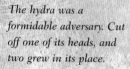

The hydra was a formidable adversary. Cut off one of its heads, and two grew in its place.

A milestone picture of colliding galaxies in Hydra, snapped by the Hubble Space Telescope. It was the telescope's 100,000th image, taken in April 2001.

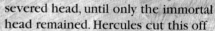
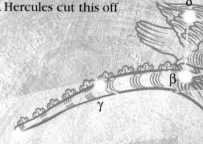

Corvus

δ

β

γ

and buried it under a rock. Then he dipped his arrows in the hydra's blood, which made them deadly.

On star maps, Hydra is depicted carrying two other constellations on its tail. They are Corvus (the Crow) and Crater (the Cup), and are important in another story. Apollo sent the crow to fetch water in a cup from a spring. On the way, it spied a fig tree and waited for days for the fruit to ripen, then gorged itself. Knowing that it would incur Apollo's wrath for such a delay, it snatched the water serpent in its claws and flew back to Apollo with it, explaining that the serpent had prevented it from filling the cup.

One solitary star

Big though the constellation is, Hydra has little to offer the casual observer. The only star of note is a second magnitude Alpha (α), named Alphard, meaning the solitary one. This is an apt description because it lies in a barren region of the heavens. It is best found by reference to the bright Regulus in Leo and Procyon in Canis Minor, with which it forms a triangle.

Factfile

★ The quadrilateral grouping of stars that forms Corvus, at the tail end of Hydra, looks good viewed through binoculars. These stars can best be located with reference to the nearby brilliant Spica in Virgo.

Hydra's second magnitude lead star Alphard is also know as Cor Hydra, meaning the hydra's heart.

Factfile

★ The Leonids—spectacular showers of meteors that rain down in November each year—were particularly heavy in 1966 and spectacular in 1999. This is because the source of the meteors, the comet Tempel-Tuttle, had just passed by.

The Egyptian sphinx might be the earthly counterpart of the heavenly constellation Leo.

Leo
The Lion

This constellation is one of the few that reasonably resembles the figure it is named after. With only a little imagination, one can see in the pattern of its stars the figure of a crouching lion. The lion constellation was revered in Ancient Egypt, for the Sun entered it at the time of the flooding of the Nile. Some consider that the Sphinx has the body of Leo.

In Greek mythology, Leo was the lion that Hercules fought with in the first of his labors—the Nemean lion. It was sent from the Moon, it was said, by Hercules' stepmother and mortal enemy Hera. The lion lived in a cave, from which it periodically emerged to prey on the local people. When Hercules came upon the lion, he attacked it with spears and arrows, but they all bounced off—its skin was invincible. So Hercules had to resort to hand-to-hand combat. After a terrible struggle, he managed to strangle the beast. Then he skinned it and made its skin into a cloak so that it would make him invincible, too.

The lion's heart

Leo is the one of the most recognizable constellations in the heavens. The curve of stars that make up the lion's head and forelimbs resembles an old-fashioned sickle used for reaping, and so is named the Sickle. At its southern end, is the brightest of Leo's stars, Regulus. It is also known as Cor Leonis, or the Lion's Heart.

Leo, and in particular the Sickle, is the apparent source of one of the most spectacular meteor showers. Called the Leonids, they may rain down

Crouching Leo is one of the most recognizable constellations.

Sickle

δ

γ

η

β
Denebola

α
Regulus

ε

A little crouching lion, Leo Minor, sits on Leo's head in this print from Urania's Mirror (1825).

on the Earth at the rate of hundreds an hour around November 17 each year.

♌ What the astrologers say

Leo is a constellation of the zodiac, lying between Cancer and Virgo. Currently, the Sun passes through Leo between August 10 and September 16.

In astrology, Leo is the fifth sign of the zodiac, covering the period from July 23 to August 22. According to astrologers, Leos can be generous and loyal, but they can also be egocentric and display a violent temper. Frequently theatrical, they are found widely among actors and film directors.

The constellation Scorpius, which now adjoins Libra and used to include its stars.

Libra
The Scales

Libra is one of the fainter constellations, that suffers by comparison with its dazzling neighbor Scorpius. But, being so located, it is an easy constellation to find. The association of Libra with scales or balance—and by extension harmony and justice—dates to Ancient Babylonian times. Then, the autumnal equinox occurred in Libra, when the days and night are of equal length, that is, in balance.

The Greeks, however, did not consider Libra to be a separate constellation at all. They saw its stars as part of Scorpius, in particular the Scorpion's claws. It was the Romans who began to identify Libra with scales again, appreciating the balance of the days and nights at the equinox. Libra is adjacent to Virgo, and the figure of Virgo is sometimes pictured holding the scales and identified as Astræa, goddess of justice.

♎ The scorpion's claws

Astronomically, there is nothing spectacular in Libra. Its two brightest stars Alpha (α) and Beta (β) are just third magnitude and make a noticeable triangle with the fainter Gamma (γ). Alpha and Beta have delightful Arabic names —Zubenelgenubi (meaning the southern claw) and Zubenelchemale (northern claw).

♎ What the astrologers say

Astræa weighs the evidence as she administers justice to the mortal world.

Libra is a constellation of the zodiac, lying between Virgo and Scorpius. Currently, the Sun passes through Libra between October 31 and November 23. In classical times, it used to be the constellation of the autumnal equinox, but this is now in Virgo.

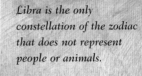

*Libra is the only
constellation of the zodiac
that does not represent
people or animals.*

β   Zubenelchemale
ε

γ

ζ   α   Zubenelgenubi

Factfile

✦ The Libran star,
Alpha, is a double
star, easily spotted
with binoculars.
✦ Beta is one of the
few green stars that
can be picked out
with the naked eye.

In astrology, Libra is the seventh sign of the zodiac,
covering the period from September 23 to October 22.
Librans are reputed to be well-balanced, as their sign
might indicate. They are quiet extroverts, but can be
indecisive. Affectionate and romantic, they are rarely
short of sexual partners.

Lyra
The Lyre

This tiny constellation, sandwiched in between Cygnus and Hercules, was sometimes pictured as a bird of prey. Indeed, the name of its brilliant star Vega, means swooping eagle in Arabic. But in mythology Lyra was the lyre. The messenger of the gods, Hermes (Mercury in Roman mythology) made the first lyre, it is said, out of the shell of a tortoise and strung it with strings made of cow gut.

Hermes gave the lyre to Apollo, whose son Orpheus learned to play it with such beauty that the wild beasts would come to listen; even the trees would follow him. Orpheus sailed with the Argonauts on their quest for the Golden Fleece. With his beautiful singing, he lulled the guardian dragon to sleep and later outsung the Sirens, whose seductive voices would have tempted the Argonauts into the sea to be drowned.

But the most famous tale of Orpheus relates to his love of the fair nymph Eurydice, whom he married. One day, while she was escaping the advances of another of Apollo's sons, Aristaeus, she was bitten by a poisonous snake and died. Orpheus felt he couldn't live without Eurydice, and descended into the Underworld to bring her back to Earth. He pleaded with Hades, the god of the Underworld, to let her go, and Hades eventually agreed, persuaded by Orpheus' beautiful music. But there was one condition: Orpheus must never look at his beloved Eurydice until they had returned to the surface. They had almost reached the gates of Hades when Orpheus looked round to see if Eurydice was still with him. Immediately, she vanished back into the

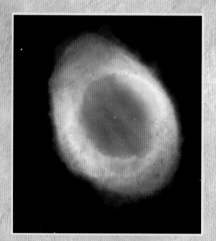

The beautiful Ring Nebula is a planetary nebula, so called because it shows up as a disc in telescopes, like planets do.

Underworld to rejoin the dead, this time for ever.

The Harp Star

The lead star in Lyra, Vega, is often called the Harp Star. Of first magnitude, it is the fifth brightest star in the heavens. It shines brilliantly high overhead in summer in the Northern Hemisphere, when it forms one corner of the conspicuous Summer Triangle of stars. The others are Deneb in Cygnus and Altair in Aquila. Lyra's other gem, between the stars Beta (β) and Gamma (γ), is a patch of gas known as the Ring Nebula (M57). Though not impressive when viewed through a small telescope, it looks exquisite through a larger one, like a formidable smoke ring. It is a ring of gas puffed out by its central dying star.

Arab astronomers identified Lyra with a bird of prey. The name Vega is derived from the Arabic meaning swooping eagle.

Apollo with the lyre Hermes gave him in exchange for the cattle Hermes stole from him.

Orion
The Mighty Hunter

Among all the constellations, Orion is outstanding. It sits on the celestial equator, which makes it equally visible to observers in both the Northern and Southern Hemispheres. And it requires only a little imagination to convert in the mind the pattern of bright stars into the figure of a mighty hunter. He has his right arm raised, ready to strike a blow with a brutal-looking club, while his left arm carries a shield. A sword hangs from his belt, marked by a diagonal composed of three bright stars.

In mythology, Orion was the son of Poseidon, the god of the sea, and Euryale, the daughter of King Minos of Crete. Poseidon gave Orion the power to walk on water. Orion went hunting with two dogs at his heels, represented in the heavens by Canis Major (the Great Dog) and Canis Minor (the Little Dog).

Orion had enormous stature and possessed prodigious strength. He was also reputed to be the handsomest of men. Unsurprisingly, many of the stories about him concern his love of beautiful women. In one, he was chasing the seven beautiful sisters known as the Pleiades. Just as he was about to catch them, Zeus intervened, turned them into doves, and later placed them among the stars. And this scenario is re-enacted in the heavens, where Orion stills pursues them.

Orion's riches

The constellation provides rich pickings for the astronomer. Its two first magnitude stars, Betelgeuse and Rigel, both supergiants, provide a nice contrast. Rigel is brilliant white while Betelgeuse is noticeably red. Another delight in Orion is the bright patch visible to

Factfile

★ Orion's star Betelgeuse is supergigantic, being one of the biggest stars known, with a diameter of some 250 million miles (400 million km), over 250 times bigger than our own Sun.

Betelgeuse is one of the biggest stars we know. If it were where the Sun is, Earth would lie inside it.

*Orion is one of
the most imposing
constellations
with beacon stars
Betelgeuse and Rigel,
that emit the energy of
60,000 suns.*

the naked eye
beneath the three
stars marking Orion's
Belt. Binoculars and
small telescopes
show that this patch
is actually a bright
nebula, a glowing
mass of gas and dust.
Telescopes are needed
to show its true
magnificence and
kaleidoscope of colors.
It is aptly named the
Great Nebula in Orion.
Astronomers know it as M42.

    The Orion Nebula is a vast star-forming
region that has been explored extensively
by the Hubble Space Telescope, which has
photographed newborn stars, some with
planetary systems in the making. Nearly
the whole of Orion is, in fact, embedded in
gas and dust. And near the star at the lower
end of Orion's Belt is another famous
nebula, the Horsehead—a dark nebula
bearing an uncanny resemblance to a
horse's head.

*Orion Nebula
can look
spectacular
through a
telescope.*

One of the many galaxies found in Pegasus. Some form groups that travel through space together.

The fine globular cluster, M15, which lies just beyond the Flying Horse's muzzle, is visible in binoculars.

Pegasus
The Winged Horse

The winged horse was a favorite concept in the ancient world. It was featured prominently in the art of the Ancient Assyrian civilization around 1500 B.C. It appeared on Egyptian coins of the same era and later on Greek coins. In Greek mythology, the flying horse, known as Pegasus, was placed among the constellations by Zeus himself.

It was Zeus' son Perseus who was responsible for Pegasus' birth. When this famous hero beheaded the snake-haired Medusa (see page 76–77), Pegasus was born from the blood spurting out of her dying body. This fabulous creature spread its wings and soared into the sky to join the nine Muses on Mount Helicon. They were Zeus' daughters, and goddesses of the arts and sciences.

The goddess Athene came to see Pegasus and tamed it with a golden bridle. She later gave the bridle to the hero Bellerophon so he could ride Pegasus when he went to fight the Chimera. This was a ferocious, fire-breathing monster, with the head of a lion, the shaggy body of a goat, and the tail of a dragon.

Having killed the Chimera, Bellerophon went on to conquer other adversaries and, proud of his achievements, flew into the sky on Pegasus to join the gods on Mount Olympus. Pegasus threw him off, and he fell back to Earth, but Pegasus went on to join the gods and serve Zeus himself by carrying his thunderbolts.

The great square

The constellation Pegasus is a dominant feature of the autumn skies in the Northern Hemisphere. It is easily recognized by its almost perfect square of four stars,

Pegasus decorates the shield of a hoplite (Greek foot soldier), 460 B.C.

which form the famed Great Square of Pegasus. The square stands out clearly in what is otherwise a relatively starless part of the heavens. All four stars in the Square are equally bright, of about the second magnitude. Only three of the stars belong to Pegasus, however: Beta (β), Alpha (α), and Epsilon (ε). The other is the lead star of the linked constellation Andromeda. With small telescopes, Epsilon, also known by its Arabic name, Enif, proves to be a double star with a faint companion.

Factfile

★ It is interesting to compare the colors of the Great Square stars, Beta, Alpha, and Epsilon. They are noticeably different —red, white, and yellow respectively.

The Great Square acts as a good signpost to fainter constellations, such as Aquarius and Pisces.

α *Andromedae*

lpheratz

β

Scheat

Great Square of Pegasus

γ *Algenib*

α

Markab

M15

ε *Enif*

Perseus
Slayer of Monsters

Perseus was one of the great legendary Greek heroes, like Hercules (see pages 62–63) and is featured in many an adventure. He is immortalized in the heavens next to his beloved Andromeda.

Perseus led an eventful life right from conception. His mother Danae, had been imprisoned by her father in an underground cell, lit only by a barred window in the ceiling. But the king of the gods, Zeus, desired her, and went to her in a shower of gold that fell as rain through the window. She became pregnant and gave birth to a boy-child, Perseus.

With a mighty blow, Perseus beheaded the snake-haired monster Medusa.

Mystified and furious, her father locked them in a chest and cast them into the sea. To make a long story short, a fisherman saved them. Later Perseus was tricked into going after the head of Medusa, one of the hideous Gorgon sisters. Medusa had snakes for hair, and her gaze would turn people into stone.

But the gods were with Perseus, giving him a shining shield, a helmet that made him invisible, a sword, and winged sandals so that he could fly. Thus equipped, he flew to where the Gorgons lived. Finding Medusa, he looked at her only as a reflection in his shield. Then he sliced off her head and flew away with it. It was on his way home that he came across Andromeda chained to the rocks, and killed the sea monster, Cetus (see pages 48–49), that was about to eat her. He later married Andromeda, who bore him six children.

In ancient art, the once-beautiful Medusa is often portrayed as a running winged demon.

⚡ The winking demon

In the sky, Perseus is depicted holding Medusa's head. The constellation lies within the Milky Way and so is a delight to sweep with binoculars. The brightest star, Alpha, is also known by its Arabic name Mirphak.

Perseus's second brightest star, Beta, is located in Medusa's head. Its Arabic name is Algol, meaning demon's head, and it has long been known as the Demon Star. For most of the time, Algol shines steadily at a magnitude of about two. But about every three days, it dims noticeably. It is a kind of variable star known as an eclipsing binary. A binary is an object that looks like one star, but is actually two, orbiting around each other. In Algol's case, one star is large and dim, the other small and bright. About every three days as we view the pair from Earth, the large, dim one covers up the small bright one, and so we see the overall brightness fall. When the dim star passes by, the former brightness returns.

French astronomer Pierre Mechain discovered M76 in 1780. It is often called the Little Dumbbell Nebula.

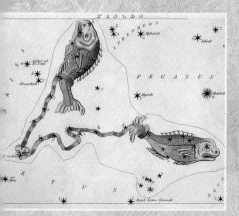

Pisces, as depicted in Urania's Mirror (1825).

Pisces
The Fishes

Perhaps surprisingly, this faint, sprawling constellation is one of the most ancient, and has always been associated with fish. As depicted in the heavens, the two fishes are swimming in opposite directions and have their tails joined by a cord. In mythology, the two fishes represent Aphrodite (Venus) and her son, Eros (Cupid). One day they had to hide in the rushes along the bank of the Euphrates River to escape the awesome dragon-headed monster Typhon. When the monster was nearly upon them, two fishes swam up and carried them away to safety, becoming immortalized in the sky as the constellation Pisces.

The goddess of beauty, Aphrodite, whom two fishes helped escape from the raging Typhon.

Watery and faint

Pisces lies in an undistinguished region of the heavens, populated by other faint "watery" constellations, such as Cetus and Aquarius. Even its brightest stars are only slightly above the fourth magnitude.

The only sure way of locating Pisces is by reference to Pegasus's famous Square. One of the two fishes lies due south of the Square, outlined by a circle of stars sometimes called the Circlet of Pisces. In this circlet is the star TX (or 19), which binoculars show to be a striking brick-red.

What the astrologers say

Pisces is a constellation of the zodiac, lying between Aquarius and Aries. Currently, the Sun passes through Pisces between March 12 and April 18. The point of intersection between the path of the Sun through the heavens (the ecliptic) and the celestial equator takes place on about March 20. It is known as the First Point

α

in Aries, because in classical times, the Sun crossed the Equator in Aries. However, because of precession—the slight gyration of the Earth's axis, the Sun now crosses in Pisces. The whole heavens have moved on since classical times.

Astrologers ignore this fact, and in astrology Pisces, the twelfth sign of the zodiac, still covers the period from February 19 to March 20. The Piscean character, astrologers say, is complex and enigmatic. Pisceans are sensitive, and often artistic and creative. They tend to fall in and out of love readily.

Factfile

✦ The vernal (spring) equinox falls on about March 20 when the Sun crosses the Equator traveling north. This signals the beginning of spring in the Northern Hemisphere.

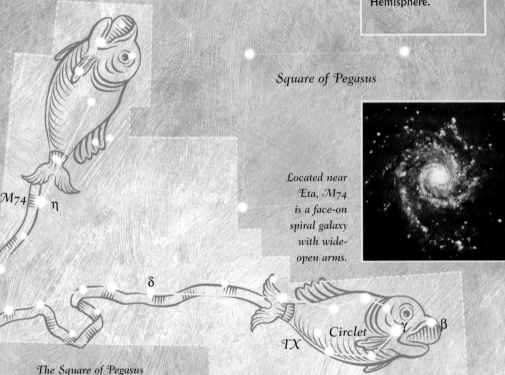

Square of Pegasus

Located near Eta, M74 is a face-on spiral galaxy with wide-open arms.

The Square of Pegasus provides the best way of locating this faint constellation.

Sagittarius
The Archer

In the heavens, we see Sagittarius as a centaur, half-man, half-horse. He is the consummate archer, keen of eye and with a deadly aim. Sagittarius dates back to Babylonian times, when the centaur was a favorite creature. But Greek historians suggest that Sagittarius was a two-legged, satyr-like beast, half-man, half-goat. A satyr is described as having a human torso, goat's legs, and a short tail. Sagittarius was supposed to be Crotus, the inventor of archery, who was fathered by the pipe-playing god Pan.

A centaur, depicted on a 2nd century sarcophagus.

Right in the center of things

Sagittarius lies in the direction of the center of our galaxy, so when you look at it, you are peering into the densest possible concentration of stars.

Possibly because of its brilliant location, Sagittarius is not the easiest of constellations to identify. It is best to first find the adjacent constellation Scorpius, which is unmistakable. The six bright stars that outline the Archer's bow and arrow then become evident. The arrow, with Gamma (γ) at the head, points menacingly at Antares, the orange-red star that marks the Scorpion's heart.

The Lagoon nebula, also known as M8, is just visible to the naked eye near the upper part of the bow. But it is best seen in binoculars or a small telescope. It makes a rough triangle with Mu (μ) and Lambda (λ). The Trifid (M20) lies slightly further north and appears as a misty patch in binoculars. A telescope is needed to show its three dark dust lanes.

Factfile

★ Sagittarius is one of the most spectacular constellations. It lies in the richest region of the Milky Way and is awash with nebulae and clusters.

★ Two of the most famous nebulae of all lie in Sagittarius: Lagoon and Trifid.

π

o

μ

M20

λ

M8

Galactic Center

δ

γ

ε

α *Rukbat*

β

Arab astronomers named Alpha, Rukbat, meaning "knee of the archer."

A glorious Hubble Space Telescope image of part of the Lagoon Nebula (M8).

♐ What the astrologers say

Sagittarius is a constellation of the zodiac, lying between Scorpius and Capricornus. Currently, the Sun passes through Sagittarius between December 18 and January 19.

In astrology, Sagittarius is the ninth sign of the zodiac, covering the period from November 22 to December 21. Sagittarians are supposed to be chatty, outgoing, and lively people, even hyperactive. Ever adventurous, they have a fiery temperament and fall in and out of love impetuously.

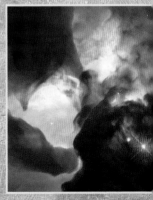

*M4 is the nearest globular
cluster, containing as
many as 60,000 stars
packed closely together.*

Scorpius
The Scorpion

Scorpius is one of the few constellation patterns that
truly resembles the figure it is meant to represent.
Only a little imagination is required to link up the
bright stars into the figure of a deadly scorpion, with its
curved tail poised ready to sting.

Scorpius is one of the oldest constellations,
recognized by the Babylonians in the Euphrates valley
7,000 years ago, and associated with their god of war.
In Ancient Greece, it was a much larger constellation
than we see now, for it included the Scorpion's claws,
which now form part of the constellation Libra. In
Greek mythology, the Scorpion was the creature that
killed the famed hunter, Orion.

Orion was the mightiest of hunters, and he knew it.
This led him, unwisely, to boast to Artemis, the goddess
of the hunt, that he could track down and kill any
creature on the face of the Earth. With this, the Earth
trembled with rage, and cracked open. A scorpion
scuttled out and stung Orion to death. With some
compassion, the gods placed Orion and the Scorpion
on opposite sides of the heavens, so that Orion sets
in the west as the Scorpion rises in the east.

Rival of Mars

The brightest star in Scorpius is Antares. It is a
huge supergiant star that is noticeably orange-red.
Its name means "rival of Mars," for its color is
similar to that of the Red Planet, Mars. Close by
Antares is a magnificent globular cluster, M4, which
is just visible to the naked eye and easily seen in
binoculars. But a telescope is needed to bring into
focus its individual stars.

♏ What the astrologers say

Scorpius is a constellation of the zodiac, lying between Libra and Sagittarius. Currently, the Sun passes through Scorpius only briefly, between November 23 and 29.

Astrologers always call this eighth sign of the zodiac Scorpio. They say it covers the period from October 23 to November 21. Scorpios, astrologers tell us, tend to be intense in everything they do, and intensely emotional. They can be easy-going, but can become testy if they are pushed too far (remember the venomous sting in the tail!). With typically warm, magnetic personalities, they make marvelous lovers.

Antares α M4

β

δ

ε

λ

Shaula

6231

ζ

Scorpius is one of the dazzling far southern constellations that northern astronomers can only glimpse.

The finest open cluster in the heavens, the Pleiades (M45), or Seven Sisters. Its hot, young stars are surrounded by nebulosity (clouds).

Factfile

★ Continuing the line through Orion's Belt and Aldebaran brings you to the Pleiades (M45), the best open cluster in the whole heavens. It is known popularly as the Seven Sisters, but not even the keenest-sighted people can make out its seven brightest stars.

Taurus
The Bull

In the heavens, Taurus depicts the front part of a bull, with a wicked red eye and horns lowered ready to charge. In mythology, the Bull represents Zeus in disguise pursuing yet another fair maiden, the beautiful Europa. Europa was the daughter of Agenor, king of Phoenicia.

Europa was playing one day on the seashore with her girl friends when she spied a beautiful white bull grazing quietly among her father's herd. Not knowing, of course, that it was really Zeus, she stroked it and climbed on its back. The beautiful bull then leapt into the sea and swam, with the now terrified Europa on his back, to Crete. There, Zeus revealed his true self and made love to her. Among the three children born of the union, one was Minos, who became king of Crete and who established bull worship at his palace at Knossos. There, too, he kept his monstrous offspring the Minotaur, half-man, half-bull, that inhabited the Labyrinth and lived on human flesh.

♉ The red eye

Taurus is a splendid constellation, close to Orion in the heavens. The three stars of Orion's Belt act as pointers to the supergiant star Aldebaran, the baleful red-orange eye of the Bull. Aldebaran is located among a V-shaped cluster of fainter stars known as the Hyades, although it actually lies much closer to us.

♉ What the astrologers say

Taurus is a constellation of the zodiac, lying between Aries and Gemini. Currently, the Sun passes through Taurus between May 14 to June 21.

In astrology, Taurus is the second sign of the zodiac, covering the period from April 20 to May 20. Taureans

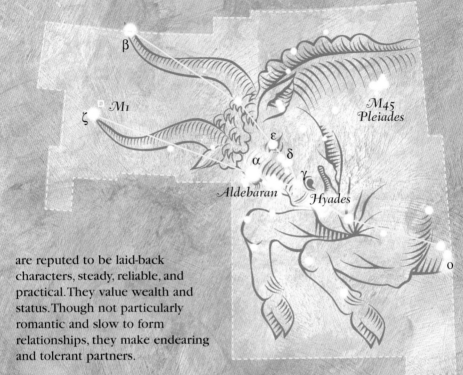

β

ζ ☐ M1

M45
Pleiades

ε
α δ
γ
Aldebaran
Hyades

o

are reputed to be laid-back
characters, steady, reliable, and
practical. They value wealth and
status. Though not particularly
romantic and slow to form
relationships, they make endearing
and tolerant partners.

The V-shaped group of
stars near Aldebaran
form an open cluster
called the Hyades.

The heart of M1, the Crab
Nebula—the remains of a
supernova seen by Chinese
astronomers in 1054.

The Minotaur was
killed by the Athenian hero
Theseus, with the help of
Minos' daughter Ariadne.

Ursa Major
The Great Bear

An old-fashioned ox-drawn plow outlined by Ursa Major's 7 main stars.

Ursa Major is a sprawling constellation that occupies a huge area of the northern skies: it rates third in size among the constellations. It is best known, not for its bear shape, but for the pattern made by its seven brightest stars. The pattern resembles both the handle and share (the part of the plow that cuts the earth) of an old-fashioned plow, and the shape of a ladle for dipping into buckets of milk or water.

The star group is called the Big Dipper—"Big" because there is a smaller dipper (the Little Dipper) close by. For much of North America and Europe, the Big Dipper is always visible in the night sky because it lies not far from the north celestial pole—it is one of the circumpolar constellations.

To the Ancient Greeks, the Great Bear represented one of Zeus' many conquests, the nymph Callisto. The daughter of Lycaon, king of Arcadia, Callisto became a companion of the virgin god Artemis and agreed to lead a life of chastity. Unfortunately for her, Zeus came by one day and became captivated by her extraordinary beauty. He transformed himself into Artemis to get close to her as she was resting. As Callisto and the false Artemis embraced, Zeus turned himself back into his usual form and promptly seduced her.

Callisto became pregnant and was banished by Artemis. She gave birth to a son, Arcas. From here on, there are several different versions of how Callisto came to be turned into a bear. One says Zeus' wife Hera did it, enraged by her husband's philandering. Later, Arcas came across the bear-like Callisto when out hunting and would have killed her. But Zeus came to her rescue and sent a

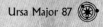

α and β act as pointers
to Polaris (Pole star)

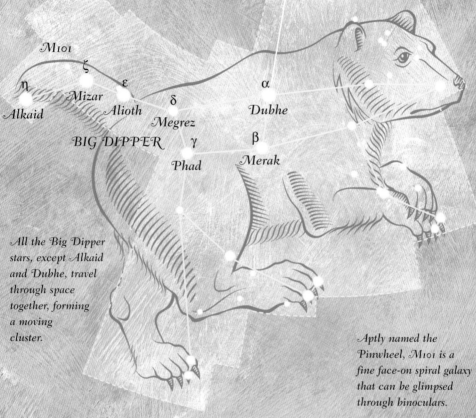

M101

ζ
η   ε
Mizar   δ   α
Alkaid   Alioth   Dubhe
Megrez   β
BIG DIPPER   γ   Merak
Phad

*All the Big Dipper
stars, except Alkaid
and Dubhe, travel
through space
together, forming
a moving
cluster.*

*Aptly named the
Pinwheel, M101 is a
fine face-on spiral galaxy
that can be glimpsed
through binoculars.*

whirlwind to whisk them into the heavens as the
constellations Ursa Major and Boötes.

The Pointers

All of the stars in the Big Dipper have names. Merak
and Dubhe in the share are known as the Pointers
because a line drawn through them points in the
direction of the Pole Star, Polaris. This fact has helped
navigators through the centuries because it is a sure
way of finding north.

Ursa Minor
The Little Bear

The Little Bear is much less conspicuous than the Great Bear. Its main star pattern imitates the shape of the Big Dipper, and is often called the Little Dipper. The tip of the Little Bear's tail is marked by the constellation's brightest star, Polaris. This is also known as the Pole Star and North Star because it lies close to the north celestial pole of the heavens, around which all the stars appear to revolve during the night.

The Greek philosopher Thales of Miletus reportedly introduced the Little Bear as a constellation in the sixth century B.C. as an aid to navigation for sailors. In mythology, the Little Bear and the Great Bear have been associated with the story of the birth of Zeus on Crete. (Other myths surround the Great Bear, see pages 86–87.)

Thales of Miletus famously predicted the eclipse of May 28, 585 B.C. that stopped a battle.

Zeus' mother Rhea had fled to a cave in Crete to escape from his father, Cronus, who had eaten all their previous children because of his fear that they might one day overthrow him. On Crete, Zeus was nursed by the nymphs Adrasteia and Ida. They made the baby a beautiful golden ball to play with that whooshed through the air like a shooting star. Cretan warriors guarded the entrance to the cave, clashing their swords and shields to drown the baby's cries so that Cronus could not hear him. When Zeus grew up, and did indeed overthrow his father, he placed Adrasteia in the sky as the Great Bear and Ida as the Little Bear.

Stars trail around Polaris in this long-exposure photo of the northern sky.

At the tip of the tail

Polaris, at the tip of the Little Bear's tail, is the Pole
Star—for the present. It lies within a degree of the
celestial north pole, and is edging ever closer—it will
be closest in the year 2095. But Polaris has not always
been the Pole Star, nor will it be in the future. Because
of a slight gyration of the Earth's spin axis, the north
celestial pole describes a circle in the heavens, coming
full circle every 26,000 years or so. In about 10,000
B.C., the star Vega in Lyra was pole star. Around
3,000 B.C., it was Thuban, the lead star in Draco.

*The location of Polaris
near the North Celestial
Pole has proved a boon
to navigators for centuries.*

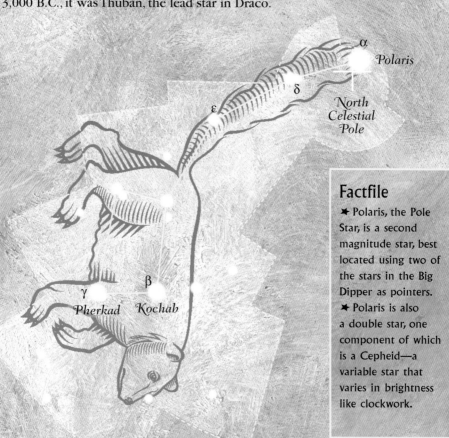

α
Polaris

δ

ε

North
Celestial
Pole

β
Kochab

γ
Pherkad

Factfile

✴ Polaris, the Pole
Star, is a second
magnitude star, best
located using two of
the stars in the Big
Dipper as pointers.
✴ Polaris is also
a double star, one
component of which
is a Cepheid—a
variable star that
varies in brightness
like clockwork.

Virgo
The Virgin

The Virgin is depicted in the heavens with wings, holding an ear of wheat. In the ancient world, she was associated with the great goddess of the harvest, since the Sun passed through the constellation around harvest time. In Ancient Babylon, she was Ishtar, reputedly a woman of immense sexual appetite who led a cult of sacred prostitution.

In Ancient Greece, Virgo was Demeter, the goddess of agriculture and fertility. By Zeus, she gave birth to a daughter, Persephone. Hades, god of the Underworld and Zeus's brother, kidnapped Persephone and took her to his underground world and made her his wife. Demeter searched far and wide for Persephone, neglecting the crops she was supposed to nurture. Eventually she learned what had happened, and Zeus persuaded Hades to return Persephone to the living Earth. But Persephone has to go back to the Underworld for part of every year because she ate some pomegranates while she was there. When she goes back each year to the Underworld, winter descends on Earth and crops die. But when she returns in the spring, the Earth becomes alive and fruitful again.

The Virgin is associated with the harvest because the Sun passes through the constellation of Virgo around harvest time.

The great cluster

Virgo is disappointing to the naked eye. Only its lead star, Spica, is bright—of the first magnitude. The main interest in Virgo lies in its deep-sky objects, in the multitude of distant galaxies visible in telescopes. These galaxies form part of an enormous group of galaxies called the Virgo cluster.

Bright star Spica marks the virgin's left hand.

Virgo cluster

ε

δ

ζ

Porrima

γ

ι

α

Spica

β

Demeter, goddess of agriculture.

♍ What the astrologers say

Virgo is a constellation of the zodiac, lying between Leo and Libra. Currently, the Sun passes through Virgo between September 10 and October 31.

In astrology, Virgo is the sixth sign of the zodiac, covering the period from August 23 to September 22. Virgos are supposed to be intelligent and modest, but they can become fussy and critical. They tend to be serious people given to worry. Overtly they are cold and unaffectionate, but this often conceals a passion within.

Chapter 4·THE WANDERING STARS

A montage of the nine "wandering stars" or planets, seen through the cameras of space probes.

Among the brightest objects in the night sky are the stars that wander against the background of the fixed stars in the constellations. Most brilliant of all is the sparkling evening star, which hangs in the western sky just after sunset on many nights of the year.

Look at these wandering stars through binoculars or a small telescope, and you will find that they are not like the other stars at all. No matter how powerful your binoculars or telescopes are, the fixed stars appear only as tiny pinpricks of light. But the wandering stars present a distinct disc. They are not remote bodies like the fixed stars, but relatively close neighbors of the Earth in space.

Two thousand years ago, the Greeks had a word for these five wandering stars—they called them planets. But we know them not by their Greek names, but by their Latin names —Mercury, Venus, Mars, Jupiter, and Saturn.

Mercury Venus Earth Mars Jupiter Saturn Uranus Neptune Pluto

SUN

The sizes of the nine planets to scale.

The Greeks thought that the planets circled around Earth. But in fact they circle around the Sun, just like Earth does—Earth is a planet, too. There are three other planets, but they are so far away that we can only see them only through a telescope. They are Uranus, Neptune, and Pluto. This makes nine planets in all.

The planets form the major part of the Sun's family in space, or the solar system. Other members of this family include the satellites, or moons, that circle around many of the planets. Then there is a swarm of "miniplanets" or asteroids, found between the orbits of Mars and Jupiter.

Comets also belong to the Sun's family. These tiny icy bodies journey in toward the Sun from the remote depths of the solar system and begin to shine only when they near the Sun and start to melt. At their brightest, they can become the most spectacular objects in the heavens. Small wonder that they caused fear and dread among the superstitious populations of past times.

The scale of the solar system is awesome. Earth is one of the inner planets that lie quite close together in the heart of the solar system. Yet it is still 93 million miles (150 million kilometers) from the Sun. The outer planets lie much further apart, separated by thousands of millions of miles. The most distant planet, Pluto, wanders at times more than 5 billion miles (7 billion kilometers) away from the Sun.

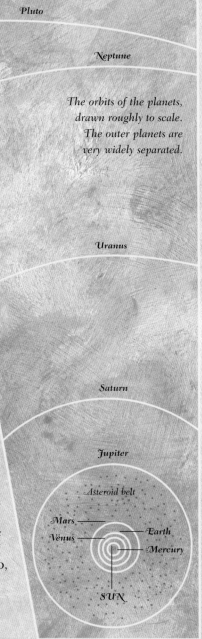

The orbits of the planets, drawn roughly to scale. The outer planets are very widely separated.

Pluto

Neptune

Uranus

Saturn

Jupiter

Asteroid belt

Mars

Venus

Earth

Mercury

SUN

The Sun

The Egyptian pharaoh Akenhaten and his wife Nefertiti worship the Sun god.

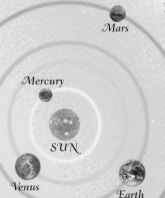

Mars

Mercury

SUN

Venus

Earth

Apollo

To us, the Sun is the most important of all the heavenly bodies, which pours light and heat onto our world. Sunlight is needed to make plants grow, and without plants for food, we and other animals could not survive. The Sun's heat warms our world and provides comfortable conditions for us and millions of other species of living things to thrive and multiply.

Small wonder, then, that ancient peoples worshiped the Sun as a powerful god. The Babylonians and the Assyrians in Mesopotamia were Sun worshipers, as were the Ancient Egyptians. To the Egyptians, the Sun god Re (or Ra) became all important. They thought that he sailed across the sky every day carrying the Sun in a boat. They pictured him with a human body and a falcon's head. The pharaohs (rulers) came to consider themselves the sons of Re. Heliopolis ("Sun City") in Lower Egypt became the center of Sun worship.

In Ancient Greece, Helios was the Sun god. It was he who drove the Sun chariot across the sky every day. He was especially worshiped in Rhodes. From about 500 B.C., however, Apollo, the god of light and purity, became increasingly identified as a Sun god. Apollo was the son of Zeus and Leto, and the twin of Artemis, who became goddess of the hunt. He was a versatile and powerful god, a celebrated musician and poet, archer, and healer. He could also reveal the future, through the Oracle at Delphi.

In later times, Sun worship dominated

A tongue of flaming gas erupts on the Sun's limb, a loop prominence of huge proportions.

the religions of three great civilizations in Central and South America—the Mayas, Incas, and Aztecs. Human sacrifice featured heavily in their rituals. The Aztecs, for example, believed that their Sun god Huitzilopochtli would die unless he was offered human blood and hearts every day.

⊛ Our neighborhood star

Astronomically, however, there is nothing special about the Sun. It is a very ordinary star, which seems much bigger and brighter than the other stars only because it is very much closer to us—only about 93 million miles (150 million kilometers) away. The distances to the stars are measured in trillions of miles.

Also, as stars go, the Sun is not particularly big or bright—there are supergiant stars that are hundreds of times bigger and tens of thousands of times brighter. In fact, astronomers class the Sun as a dwarf—a yellow dwarf because the light it gives off is a yellowish color.

During an eclipse, the Moon moves across the face of the Sun and turns day into night.

Covering up on the beach helps prevent sunburn from the Sun's invisible ultraviolet rays.

The Moon

The Moon is Earth's constant companion and closest neighbor in space. It is the only other world that humans have set foot on and explored—so far. Apollo astronauts made six landings on the Moon between 1969 and 1972.

The Moon is Earth's only natural satellite, circling around Earth once a month. During this time we see it appear to change in shape—go through its phases—from slim crescent to full circle and back again. It lightens our world by night, just as the Sun lightens our world by day. Like the Sun, the Moon was widely worshiped in the ancient world. In early Greek mythology, the Moon goddess was Selene, also called Mene. She was the sister of the Sun god Helios. Every evening, as Helios finished his journey across the daytime sky, Selene started hers in the nighttime sky. Later, the goddess Artemis (Diana in Roman mythology) took on the role of a Moon goddess. She was a divinity of the light like her brother Apollo, who became a Sun god. Artemis was also goddess of the hunt and wild animals.

The Greek goddess Artemis was the sister of the Sun god, Apollo.

One of the Moon's most prominent craters, Copernicus.

The full Moon, 14 days old. In the south, brilliant rays fan out from the crater Tycho.

The crescent Moon, only a few days old, displaying the circular Sea of Crises.

Artemis was the ultimate virgin goddess who gathered around her a band of maidens vowed to chastity. Those who lost their virginity even by trickery (like Callisto) were banished in disgrace. And woe betide any mortal Peeping Tom, such as the lovelorn Actaeon, who happened upon Artemis bathing naked. Artemis changed Actaeon into a stag and then set her hounds upon him, that tore him limb from limb and then gobbled him up.

⊛ A dead world

The Moon is a rocky body like Earth, but is a very different world. It is much smaller than Earth—only about a third of the size across. Being so small, it has a small mass and low gravity (only one-sixth Earth's gravity). As a result, the Moon could not retain any atmosphere. With no atmosphere, there are no clouds on the Moon, no rain, and no blue sky. And temperatures are extreme. They soar to more than 250°F (120°C) during the lunar day and drop to below -240°F (-150°C) during the lunar night, both of which are roughly two Earth-weeks long.

The Moon circles around Earth in just 27.3 days, and it also spins on its own axis in the same period. As a result of these motions, the Moon always presents the same face towards us—the nearside. Even with our eyes we can make out two different kinds of regions on the nearside, dark and light. Through binoculars and telescopes we can see that the darker regions are great plains, which early astronomers called seas. And the lighter regions are highlands. We can see craters everywhere, some measuring hundreds of miles across.

Factfile

➤ Diameter at equator: 2,160 miles (3,476km)
➤ Mass: 1/81 Earth's mass
➤ Average distance from Earth: 239,000 miles (384,000km) Circles around Earth in: 27.3 days
➤ Spins on its axis in: 27.3 days
➤ Goes through phases in: 29.5 days

Man on the Moon. Astronaut James Irwin at the Apollo 15 landing site in 1971.

Mercury

Mercury is the planet closest to the Sun, and the one that travels the fastest. Because it never strays far from the Sun in the sky, it is not an easy planet to spot. It can either be seen low down near the eastern horizon before dawn or low down near the western horizon after sunset. At its brightest, it is slightly more brilliant than Sirius, the brightest star.

The barren surface of Mercury, photographed in 1974 by the space probe Mariner 10.

Ancient astronomers were familiar with Mercury when it appeared in the morning and the evening, but they did not realize it was the same body. When they saw it as a morning star, they called it Mercury, but called it Apollo as an evening star.

Mercury is an appropriate name for this fast-moving planet, for Mercury was the fleet-footed messenger of the gods in Roman mythology. He is usually depicted as a lithe, handsome youth with a winged helmet (petasus) and winged sandals so that he could fly swiftly to do the gods' bidding. In one hand he carries a magic wand called the caduceus, which has wings at the top and snakes twined round it. Mercury was known as Hermes in Greek mythology and was the son of Zeus and Maia, a daughter of Atlas—the Titan condemned forever to hold the heavens on his shoulders. On the very day he was born, Hermes displayed his talents for mischief and invention. He stole cattle from

Mars

MERCURY

Sun

Venus

Earth

Apollo and constructed a new musical instrument, the lyre. He subsequently presented it to Apollo to make amends for the theft of his cattle.

● Another moon

Mercury is the smallest of the terrestrial, or Earth-like planets. It is less than half as big across as Earth. As a result it has a low mass and low gravity and has been unable to hold onto any atmosphere. Being close to the Sun, we would expect it to be hot, and it is. Parts of the planet facing the Sun reach temperatures up to 800°F (430°C)—which is hot enough to melt lead. But at the same time, temperatures on the opposite side of Mercury facing away from the Sun plummet to -290°F (-180°C).

Photographs returned by the space probe Mariner 10 show that Mercury is covered with craters, and looks much like some parts of the Moon. These craters were dug out by meteorites billions of years ago. A particularly big one created a huge basin called the Caloris Basin. Some 800 miles (1,300 kilometers) across, it is the most prominent feature on the planet.

Factfile

★ Diameter at equator: 3,031 miles (4,878km)

★ Average distance from the Sun: 36 million miles (58 million km)

★ Mass (Earth=1): 0.06

★ Spins on its axis in: 58.7 days

★ Circles around the Sun in: 88 days

★ Number of moons: 0

Mercury's heavily cratered surface makes the planet look much like the Moon.

Mercury (Hermes), the fleet-footed messenger of the gods.

Venus

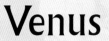

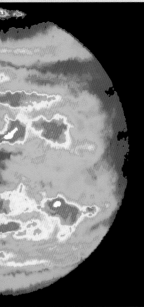

Clouds in Venus's thick atmosphere show up in this false-color image taken by the Galileo probe.

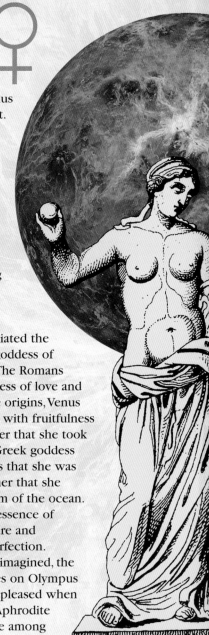

O f all the planets, Venus is the easiest to spot. It is by far the brightest —far brighter than any of the stars. It hangs in the early evening sky on many nights of the year, in the west just after sunset. It is the evening star. At other times of the year, early risers may also see Venus as the morning star, hanging in the east as the sky brightens just before sunrise.

The Babylonians associated the planet with Ishtar, their goddess of beauty, fertility, and war. The Romans named it after their goddess of love and beauty, Venus. Of obscure origins, Venus was originally connected with fruitfulness and crops. It was only later that she took on the attributes of the Greek goddess Aphrodite. One story tells that she was a daughter of Zeus, another that she was born from the foam of the ocean. Aphrodite was the essence of beauty, whose figure and features were perfection.

As might be imagined, the other goddesses on Olympus were not well pleased when the beautiful Aphrodite took her place among them. Her main rivals

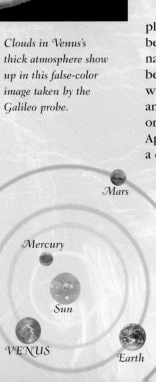

Mars

Mercury

Sun

VENUS

Earth

were Hera and Athene. One day all three were arguing about which one was the fairest, when Zeus intervened and said that a mortal must make the decision.

Paris, the son of King Priam of Troy, was made to choose. To help him make up his mind, Hera promised that he would be lord of all Asia if he should judge her the most beautiful; Athene, that he would be invincible in battle. Aphrodite promised that she would help him win the most beautiful mortal woman. There was no contest—he chose Aphrodite. And with Aphrodite's help, he carried off the beautiful Helen, wife of Menelaus, the king of Sparta, a powerful city-state in Greece. His refusal to hand her back sparked the epic Trojan War, in which the Greeks were ultimately successful by using the Trojan horse.

Aphrodite's beauty stirred all of the gods, but strangely, she took the ugliest among them, Hephaestus (Vulcan in Roman mythology) as a husband. But it was inevitable that she should console herself with some of the more fetching of the gods, including Ares (Mars) and Hermes (Mercury). She also loved two mortals, Adonis and Anchises. From the union with the latter, she gave birth to Aeneas, one of the few to survive the fall of Troy. His wanderings thereafter were the subject of one of the world's great literary epics, the *Aeneid*, written by the Roman poet Virgil.

Under the clouds, Venus's surface has been shaped by lava flows from huge volcanoes.

Venus, one story tells, was born from the foam of the oceans.

Aphrodite

Factfile

* Diameter at equator:
7,521 miles (12,104km)
* Average distance
from the Sun:
67 million miles
(108 million km)
* Mass (Earth=1): 0.8
* Spins on its axis in:
243 days
* Circles around the
Sun in: 225 days
* Number of
moons: 0

The Greenhouse Planet

Venus is a near twin of Earth in size, is made up of rock, and has an atmosphere. But the two bodies have little else in common. For one thing, the temperature on Venus reaches an amazing 900°F (480°C). It is hotter than Mercury, even though it is nearly twice as far away from the Sun. The atmosphere is responsible for this, acting like a greenhouse to trap the Sun's heat. As on Earth, the major cause of the greenhouse effect is carbon dioxide, the predominant gas in Venus's atmosphere. The effect is magnified by the sheer bulk of the atmosphere, whose pressure is ninety times that of the Earth's atmosphere.

Under the veil

Thick clouds blanket Venus's atmosphere. But they are not like the innocuous water-droplet clouds we have on Earth. Venus's clouds are made up of droplets of sulfuric acid. The dense cloud layers hide the surface from view

An overhead view of
Venus's Sapas Mons
volcano, about a mile
(1.6km) high.

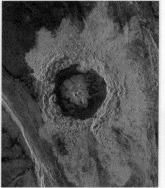

One of Venus's few
large meteoric craters
(the 43-mile (70-km)
wide Dickinson.

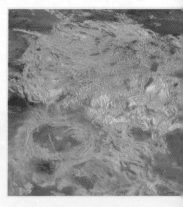

Alpha Regio is one of the
prominent upland regions
on Venus, some 800 miles
(1,300km) across.

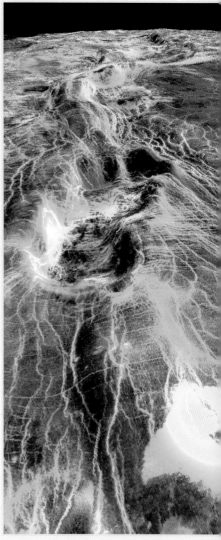

when we look at the planet in a telescope. But we know what Venus looks like because orbiting space probes like Magellan have used radar to see through the clouds.

As expected, there are no oceans on Venus like there are on Earth. Any there might have been would have boiled away long ago because of the intense heat. Most of Venus's surface consists of low-lying plains, with a few highland areas dotted about. The two largest highlands can be thought of as continents. One, named Aphrodite Terra, lies near the equator and is about the size of Africa. The other, Ishtar Terra, lies further north and is about the size of Australia. Ishtar Terra is dominated by a vast plateau named Lakshmi Planum and the highest mountains on Venus, named Maxwell Montes.

Volcanoes, volcanoes everywhere

The surface of Venus has been shaped almost entirely by volcanoes. They are found everywhere and have erupted repeatedly. The runny lava they have spewed out has been able to travel for tens and even hundreds of miles, creating Venus's relatively smooth rolling plains.

The volcanic landscape on Venus shows some unique features, not seen on Earth or anywhere else in the solar system. They include flat "pancake" domes, crown-like structures called coronae, and networks of fine cracks that look like spiders' webs, which are appropriately called arachnoids.

Sedna Planitia is an extensive lowland region, criss-crossed by long fractures in the crust.

Earth

The planet we live on is the largest of the rocky, or terrestrial (Earth-like) planets. It is unique among these and indeed all the planets because it is a cradle of life. It has just the right conditions to allow an abundance of life to survive and thrive. Such conditions do not exist on any other planet— at least in our part of the universe.

Earth was personified as a deity in both Egyptian and Greek mythologies. To the Egyptians he was the god Geb, whose sister and wife was the sky goddess Nut. It was her star-studded body that was held over Geb by Shu, a kind of Egyptian Atlas.

The Greeks conceived Earth as a deep-breasted goddess, Gaea. She emerged out of the primeval emptiness they called Chaos. She bore Uranus, the god of the heavens. The Earth and the heavens—the universe—was created, but not peopled. So Gaea lay with Uranus to beget the first race of the gods, the Titans, and the one-eyed Cyclopes. She is also credited as giving birth to the human race of mortal men and women. Gaea had the gift of prophesy and was venerated at the Oracle at Delphi before she was supplanted by Zeus.

Earth has only one natural satellite, the Moon, but swarms of artificial ones.

In Egyptian mythology, the sky goddess Nut spanned her husband, the Earth god Geb.

⊛ Continents adrift

Earth has the typical layered structure of all the rocky planets. It has a thin, hard crust, 70 percent of which is hidden by the water of the oceans. The crust overlays a deep rocky

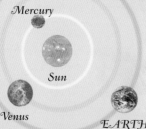

Mars

Mercury

Sun

Venus

EARTH

layer known as the mantle. At the Earth's center is a large core of iron and nickel, the outer part of which is liquid. Currents eddying inside the liquid core create electric currents and magnetism, turning Earth into a giant magnet, with a magnetic field extending far out into space.

The outer crust of Earth is not solid like an eggshell, but is split into sections known as plates. These plates are in constant motion over the face of the Earth, causing the oceans to widen and the continents to drift slowly apart. Movements along the plate boundaries set off earthquakes and trigger volcanic eruptions.

The changing landscape

Earthquakes and volcanoes bring about sudden changes to the Earth, but more subtle forces of change are also at work. They are the forces that cause erosion, the gradual wearing away of the landscape. Weathering —the action of the weather—is a powerful erosive force. The Sun's heat, frost, and the sandblasting effect of dust-laden wind all take their toll. Flowing water is also destructive, cutting into and dissolving rocks and depositing the debris elsewhere.

Factfile
✷ Diameter at equator: 7,927 miles (12,756km)
✷ Average distance from the Sun: 93 million miles (150 million km)
✷ Mass : 1
✷ Spins on its axis in: 23.94 hours
✷ Circles around the Sun in: 365.25 days
✷ Number of moons: 1

Erupting volcanoes (below left) show that Earth is still very much alive.

The Colorado River has been eroding the Grand Canyon (below) for millions of years.

Mars

Factfile

★ Diameter at
equator: 4,221 miles
(6,792 km)
★ Average distance
from the Sun:
142 million miles
(228 million km)
★ Mass
(Earth=1): 0.11
★ Spins on its axis
in: 24.63 hours
★ Circles around the
Sun in: 687 days
★ Number of
moons: 2

Mars cuts a distinctive figure
in the night sky because
it boasts a fiery reddish-
orange color, which has
earned it the title of the
Red Planet. Like Venus,
Mars is a neighbor of Earth
in space, approaching
at times to within 35
million miles (56 million
kilometers). But it is much
smaller than Venus, and
therefore does not shine as
brightly. Nevertheless, it still
outshines all the stars in the
sky when it gets closest to us
and most brilliant.

Mars is named after the Roman
god of war, probably because its
reddish hue was reminiscent of fire and
bloodshed. But Mars was originally a gentler
god, of agriculture and fruitfulness. When Mars
became a god of war, he took on the attributes and
legends of the Greek god of war, Ares.

Ares was the son of Zeus and Hera, but was
not popular among the other gods because of
his brutality on the battlefield. He took the
goddess of love and beauty, Aphrodite, as
his wife, and she bore him three
children—two sons (Phobos
and Deimos) and a daughter
(Harmonia). Phobos (Fear),
and Deimos (Terror) accompanied
Ares into battle, along with
his sister Eris (Strife). Mars's

MARS

Mercury

Sun

Venus

Earth

two moons are named after Ares's two sons.

⬢ The myth of life

A century ago, many people believed that Mars was inhabited by an intelligent race of Martians. They pointed out similarities between Mars and Earth—it has ice caps; its day is only slightly longer than our own; "waves of darkening" sweep across the planet at times, which could signal the growth of vegetation. The Martian concept was triggered by the Italian astronomer Giovanni Schiaparelli in 1877, who reported seeing "canals" on the planet. People interpreted this as meaning artificial waterways, and thought intelligent beings lived there. One particular enthusiast was the American astronomer Percival Lowell, who built an observatory at Flagstaff, Arizona, to study Mars more closely. He conjured up a pathetic picture of a dying breed of Martians, desperately digging canals to channel precious water away from the polar icecaps. A good story, but only that. Space probes have found that conditions on Mars are too extreme for any kind of life to survive.

Mars (Ares) the god of war, was feared for his brutality.

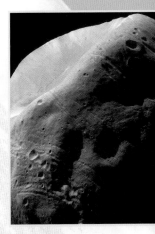

Phobos is the largest of the two moons of Mars. Some 16 miles (26km) across, it is twice the size of Deimos.

One of Lowell's drawings, showing a network of Martian canals.

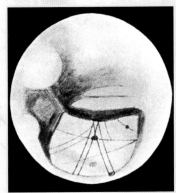

The Red Planet

Distinctly red in the sky, Mars proves to be red on the ground, too. Space probes that have orbited the planet and landed on its surface attest to this. Much of the planet consists of barren, desert-like regions, covered with rust-colored sand and rust-colored rocks. Even the Martian sky takes on a pale reddish hue because of the presence of fine particles kicked up by the winds that quite frequently blow across the planet.

Sometimes the winds blow with gale-force velocity and trigger planet-wide dust storms that obscure the surface from view. That strong winds should blow on Mars is perhaps surprising, given that the Martian atmosphere is so slight—the atmospheric pressure is only one-hundredth of that on Earth.

The vast desert regions of Mars are found mainly in the northern hemisphere. Astronomers have speculated that they may once have been vast oceans. Martian deserts are not baking hot, like many of Earth's. Temperatures on Mars are generally well below freezing for most of the time. The southern hemisphere is much more rugged and pockmarked with craters where it has been bombarded with meteorites in the distant past.

Martian winds whip up dust into the atmosphere to create dust storms.

This view of Mars shows two volcanoes in Tharsis at the top and, on the left, the great scar of Mariner Valley.

⊛ Volcanoes and valleys

But it is near the equator that the most awesome natural features on Mars can be found. In a region named Tharsis, the surface bulges dramatically and is topped by a line of three massive volcanoes, towering up to nine miles (fifteen kilometers) high. But they are dwarfed by a fourth to the west, appropriately named Olympus Mons, or Mount Olympus. Some 400 miles

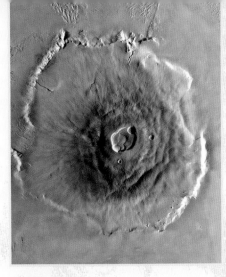

The magnificent Olympus Mons, the most massive volcano we know in the whole solar system.

(600 kilometers) across at the base, it soars to a height of over fifteen miles (twenty-five kilometers), or three times the height of Earth's highest peak, Mount Everest.

Near the equator to the east of Tharsis, an enormous system of canyons and valleys begins. It is named Mariner Valley (Valles Marineris) after the Mariner probe that discovered it. It is popularly called Mars's Grand Canyon, but with a length of more than 2,500 miles (4,000 kilometers) it is at least ten times longer than the Grand Canyon in Arizona.

⊛ Water, water

Clouds in the atmosphere and ice caps at the north and south poles testify that there is water on Mars, at least in small amounts. And scientists are now convinced that there was abundant water on the planet long ago, when the climate was warmer than it is now. There might still be much water locked as ice in the ground, like permafrost on the Arctic tundra on Earth.

The tiny Sojourner rover inches its way over the Martian surface in 1997.

Jupiter

Factfile

* Diameter at equator: 88,900 miles (143,000km)
* Average distance from the Sun: 484 million miles (778 million km)
* Mass (Earth=1): 318
* Spins on its axis in: 9.93 hours
* Circles around the Sun in: 11.9 years
* Number of moons: 16

When you see a planet shining brilliant white in the dead of night, not in the twilight of dawn or dusk, then it is bound to be Jupiter. (The brilliant white planet of the twilight is Venus.) Jupiter is visible for much of the year, shining consistently bright. The only planet to rival it in brilliance in the darkness of night is Mars. But Mars can easily be recognized by its distinctive reddish hue.

Jupiter takes about twelve years to circle around the Sun, and like all the other planets, it travels against a background of the twelve constellations of the zodiac (see pages 22–23). This means, roughly speaking, that it passes through one of the constellations every year. In late 2001, Jupiter was to be found in Gemini, crossing into Cancer in 2002. By 2010, it will be in Aquarius.

By Jove

Jupiter was the king among the Roman gods. He was also called Jovis, from which our exclamation "By Jove!" comes. He was the embodiment of the Greek god Zeus and like him the god of the sky and the elements, often pictured hurling thunderbolts. The mythology surrounding the top god in the whole pantheon is quite extensive. Zeus was born the son of Cronus and Rhea, who spirited him away after birth to prevent him from being swallowed by his father (see page 115).

As a man, he overthrew Cronus and divided up the world with his brothers Poseidon and Hades. To Poseidon he gave

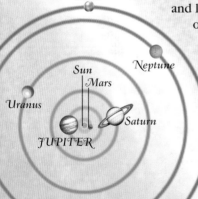

Pluto

Sun

Neptune

Mars

Uranus

Saturn

JUPITER

control of the sea, and to Hades the Underworld. He kept the sky and the heavens for himself, and dwelt on Mount Olympus. Zeus was the ruler of the gods and of mortals. He was omnipotent—he could see everything and he knew everything.

The love life of Zeus was prodigious. First he married Metis, the goddess of wisdom. But it was prophesied that his first child would be wiser than its father, so he swallowed his wife and unborn child. But before long he developed an intolerable headache, that Hephaestus cured by splitting his skull. Out of the gaping wound leapt his offspring, not as a babe but fully grown, clad in armor and brandishing a javelin. She was Athena, the goddess of wisdom and war.

His other wives included Themis and, most notably, Hera. When married to Hera, he carried on a profusion of illicit affairs both with goddesses and mortal women, often transforming himself into other people or animals to pursue his seductions.

Jupiter presided over the other gods on Mount Olympus.

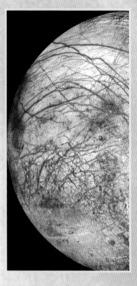

Jupiter has a swarm of moons circling around it—this one is Europa.

Through a telescope, colored bands are clearly visible in Jupiter's atmosphere.

Great swirls of clouds surround Jupiter's Red Spot.

NASA's Galileo probe spots a volcanic eruption and lava flow on Io.

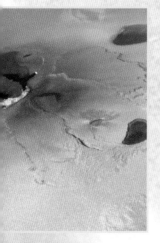

King of the Planets

That Jupiter should be named after the king of the gods is most appropriate, for it is truly the king among the planets. With a diameter eleven times that of Earth, it is bigger than all the other planets put together. In the solar system, only the Sun is bigger. And if Jupiter had about fifty times more mass than it has now, it too would have become a star—another Sun.

And, like the Sun, Jupiter is made up mainly of gas. It is not a rocky body like Earth and the other three inner planets. Astronomers call it a gas giant, along with Saturn, Uranus, and Neptune. Jupiter has a deep atmosphere containing mainly hydrogen and helium. At the bottom of the atmosphere, the pressure is so high that it turns the hydrogen gas into liquid, creating a deep liquid hydrogen ocean. In turn, at the bottom of this ocean, the now fantastic pressure converts the hydrogen into a kind of liquid metal, rather like mercury. Only at the center of Jupiter is there a tiny core of rock.

The stormy atmosphere

In a telescope, the face of Jupiter presents a colorful spectacle. It is crossed by pale and dark bands that are mainly reddish-brown in color. Astronomers called the pale bands zones and the darker ones belts. Space probes have shown that the colorful zones and belts are layers of clouds. They are stretched out into parallel bands by the planet's rapid rotation—it spins around in space faster than any other planet.

The winds in the belts and zones blow in different directions, causing violent turbulence. The swirls and eddies thus created show up clearly on Jupiter's face. Powerful storms break out all over the planet, revealed

The icy surface of Europa is constantly changing as it breaks up and then refreezes.

as pale and colored spots. The biggest storm region, and a permanent feature, is the Great Red Spot, a superhurricane three times as big across as Earth.

Galileo's moons

When Galileo turned his telescope on Jupiter in 1610, he spotted four tiny moons circling round it. These Galilean moons are Io, Europa, Ganymede, and Callisto. With a diameter of 3,278 miles (5,276 kilometers), Ganymede is the largest moon in the solar system, larger even than the planet Mercury. Io is the most interesting moon, however, because of its many active volcanoes, which spew out molten sulphur, not rock, and give the moon its garish yellow-orange color. Since Galileo's time at least twelve other moons have been discovered, all very much smaller.

Young groove-like features cut into older terrain on Ganymede's surface.

The four Galilean moons, in order of size. From the left, Ganymede, Callisto, Io, and Europa.

Factfile

★ Diameter at equator: 74,900 miles (120,500 km)
★ Average distance from the Sun: 887 million miles (1,427 million km)
★ Mass (Earth=1): 95
★ Spins on its axis in: 10.66 hours
★ Circles around the Sun in: 29.5 years
★ Number of moons: 18+

Saturn

Saturn is the most distant planet that we can easily see with the naked eye. It is nearly twice as far away as Jupiter and naturally never becomes as bright. But at its most brilliant, it still outshines all the stars in the sky except Sirius and Canopus. Unlike brilliant white Jupiter, it shines with a distinctly yellow light.

Saturn is much slower moving than Jupiter, taking nearly thirty years to circle around the Sun. It thus spends on average two years or so traveling through each of the twelve constellations. It spent most of 2001/2002 traveling through Taurus (incidentally, pairing in the heavens with Jupiter, which was in the adjacent Gemini). By 2010, it will be in Virgo.

Saturn (Cronus) was god of the harvest.

God of the harvest

The Romans worshiped Saturn as god of the harvest, and as to be expected, he was one of the most important of the gods. One of the most riotous festivals in Roman times was the Saturnalia, which honored the god and celebrated the harvest. For the week of December 17, all work and business ceased, slaves were given temporary freedom, and presents were exchanged. We see this festival paralleled today in Christmas and New Year celebrations.

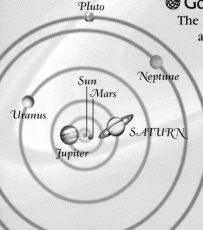

As ever, the Romans borrowed the persona of Saturn from the Greeks, who called him Cronus. He was the youngest son of Uranus, god of heaven, and Gaea, goddess of the Earth. He attacked Uranus to set free his siblings, who were imprisoned (see page 119), and married his sister, Rhea. She gave birth to six children, but Cronus ate the first five because an oracle had foretold that one of them would overthrow him. When Rhea gave birth to her sixth child, Zeus, she hid him away but wrapped his clothes around a stone and presented it to Cronus, who swallowed it immediately.

This false-color image shows variations in the make-up of Saturn's rings.

Zeus was raised by the nymphs Adrasteia and Ida, daughters of the king of Crete (see page 88). When Zeus reached manhood, he vowed vengeance on his father by getting him to swallow a potion that made him vomit up the stone, and the five children that he had swallowed—three girls (Hestia, Hera, and Demeter) and two boys (Hades and Poseidon), who all became gods and goddesses. As the oracle had predicted, Zeus took over Cronus's throne and banished him to the ends of the Earth.

A montage of Saturn and five of its moons. They are, from the left, Rhea, Enceladus, Dione, Tethys, and Mimas.

The Ringed Planet

A narrow F ring lies outside the ring system we can see from Earth.

Turbulent, stormy regions in Saturn's atmosphere show up as waves and oval spots.

When Galileo first looked at Saturn in his telescope in 1610, he noticed that there was something odd about it. It appeared to have "ears," or projections on either side. Only later, when telescopes became more powerful, were these "ears" resolved into a set of shining rings, circling the planet's equator.

In a telescope we see three main rings, designated A, B, and C from the outside in. The B ring is the densest and most brilliant, the C ring the dimmest and most transparent. Between the A and B rings is a dark gap, called the Cassini division. There is also a smaller gap, the Encke division, near the outside of the A ring. From one edge of the A ring to the other, the rings span a distance of some 170,000 miles (270,000 kilometers).

When the Voyager 1 and 2 space probes visited the planet (in 1980 and 1981), they discovered additional rings inside and outside the known system, but they were much narrower and fainter. The probes also discovered many small new moons, three of which proved particularly interesting. One (Atlas) circles just outside the A ring, and two (Prometheus and Pandora) circle on either side of the newly discovered F ring. Astronomers believe that these tiny moons somehow keep the icy ring particles in place and call them shepherd moons, since they "herd" the ring particles in much the same way as a shepherd herds his sheep.

🪐 The body of Saturn

Saturn is somewhat smaller than Jupiter but seems to have an almost identical make-up. Under a hydrogen/helium atmosphere, there is a deep ocean of liquid hydrogen. It sits on top of a liquid metallic hydrogen layer, with a small rocky core in the center. The atmosphere is banded like Jupiter's but not so distinctly—it is overall much hazier. There seem only to be a few small storm centers, and certainly nothing rivaling Jupiter's Great Red Spot.

Saturn's many moons

Saturn has more moons even than Jupiter—eighteen have had their orbits confirmed, and several others have been spotted from time to time. The larger ones seem to be made up of a mixture of water, ice, and rock. The biggest one, Titan, is unusual. With a diameter of 3,200 miles (5,150 kilometers), it is the second largest moon in the solar system, after Jupiter's Ganymede. But unlike all other moons, it has a thick atmosphere of mainly nitrogen and methane. At the temperature of Titan, about -290°F (-180°C), methane might take on the role water has on Earth. And there might be methane lakes and snow on the moon, and drops of methane raining down from the clouds.

Because of its tilted axis, our views of Saturn change over the years as it travels in orbit around the Sun.

Titan is unique among moons, because it has a thick atmosphere, containing mainly nitrogen gases.

Many craters scar the surface of Dione. The largest approach 60 miles (100km) across.

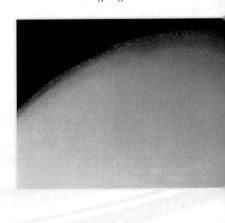

Uranus

Factfile

* Diameter at equator: 31,800 miles (51,100km)
* Average distance from the Sun: 1,787 million miles (2,875 million km)
* Mass (Earth=1): 15
* Spins on its axis in: 17.24 hours
* Circles around the Sun in: 84 years
* Number of moons: 18+

Saturn was the most distant planet known to ancient astronomers. And no one suspected that there might be other planets—until 1781. In March of that year in England, a German-born musician-turned astronomer named William Herschel spotted an object in the constellation Gemini that he first thought was a comet. But it wasn't—it was a new planet, which came to be called Uranus.

Uranus can, just, be seen with the naked eye, if you know exactly where to look. It is very slow moving among the constellations for it takes 84 years to circle the Sun. Late 2001 found it in Capricornus, and 2002 will see it crossing into Aquarius, where it will remain for nearly seven years.

The Georgian planet

Since the ancients did not know about Uranus, the planet has no mythology as such. Herschel called it the Georgian planet, in honor of his patron, George III. The German astronomer Johann Bode suggested the name Uranus, which fitted in well with the other planets' names. In Greek mythology, Uranus was god of the heavens and one of the oldest gods, whom the Earth goddess Gaea bore and then married. They fathered the Titans, the Cyclopes, and the Hecatoncheires,

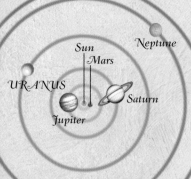

Pluto
Neptune
Sun
Mars
URANUS
Saturn
Jupiter

George III

monsters with 100 hands and 50 heads.

Uranus was so ashamed of his offspring that he locked them away in the depths of the Earth. But Gaea conspired with her last-born son, Cronus, to attack Uranus. He castrated his father with a sickle and threw his genitals into the foaming sea. From the drops of blood that spattered Gaea, the Giants and the Furies were born, and from the foam on the sea rose Aphrodite.

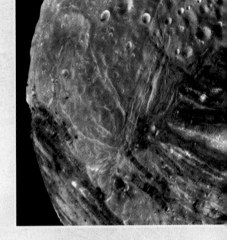

⬡ Topsy turvy

Uranus is the third largest planet, about four times bigger across than Earth. It is one of the four gas giants, with a very deep atmosphere, an even deeper ocean containing water, ammonia, and methane, and a central core of rock. Its face is almost featureless.

The strangest thing about Uranus is that its axis is tipped right over in relation to its path around the Sun. All the other planets spin around their axis in a more or less upright position, but Uranus spins on its side. Like Saturn, Uranus has a set of rings circling the equator, but they are made up of dark particles and are very faint.

Uranus has a family of moons rivaling Saturn in number. The weirdest is Miranda, which has a patchwork of contrasting features. Astronomers think that long ago it might have been knocked to pieces in a collision and then reformed with the pieces stuck in different positions.

Miranda's crazy surface, unlike that of any other body in the solar system.

Few details can be seen in Uranus's atmosphere. In false color, a slight haze (dark region) shows up over the south pole.

Neptune's Dark Spot was a stormy region spied by Voyager 2 in 1989.

French mathematician Urbain Jean Joseph Leverrier (1811–77).

Neptune

After Herschel's discovery of Uranus, astronomers found that it deviated somewhat from its predicted orbit around the Sun. So they suspected that another unknown planet must be affecting it gravitationally. Two mathematicians—John Adams in England and Urbain Leverrier in France—worked out where this planet should be. And in September 1846 a German astronomer, Johann Galle found it.

Neptune is much farther away than Uranus and is entirely beyond the reach of the naked eye. But it can be spotted through binoculars if you know where to look. Since it takes about 165 years to orbit the Sun, it hardly seems to move in the sky at all as the years go by. It has been traveling in Capricornus since 1997 and will remain there until about 2010.

The sea god

Galle proposed the name Janus for the eighth planet. Leverrier first suggested Neptune, then changed his mind and proposed to name it after himself! But Neptune soon came to be accepted. Again, strictly speaking, Neptune has no mythology because it was unknown to the ancients.

In Roman mythology, Neptune was the god of the sea. He was borrowed from Greek mythology, in which the sea god was

Pluto

Sun
Mars
NEPTUNE
Uranus
Jupiter
Saturn

called Poseidon. Poseidon was one of the offspring of the Titan Cronus and his sister-wife Rhea and was the brother of Zeus and Hades. One tale about Poseidon concerns a contest with the goddess Athena. The Greeks of Attica wanted to name their chief city after whoever gave mankind the most useful object. Poseidon created the horse for them, but they favored Athena's gift—the olive tree. That is how the city of Athens got its name.

⚛ Twin world

In size and make-up, Neptune is a near twin of Uranus, being only fractionally smaller. However, it is a much more interesting planet to look at. It has vague bands like Jupiter and Saturn, and spots that reveal where storms are breaking out in its atmosphere. Voyager 2, the only probe to visit Neptune (1989), spotted a particularly large storm region, which was named the Great Dark Spot. It was edged with fluffy clouds, much like the cirrus clouds we call mares' tails on Earth. Winds blowing around the spot were clocked at speeds approaching 1,200 mph (2,000 kph)—the fastest ever seen in the solar system.

Like Uranus, Neptune has rings around it—two bright narrow ones and two faint wide ones. They are far too faint to be seen from Earth. Among Neptune's moons, Triton is the largest. It is a deep-frozen world that has icy geysers erupting over its surface.

Neptune (Poseidon) is nearly always pictured holding a three-pronged spear, or trident.

Factfile

✶ Diameter at equator: 30,800 miles (49,500km)

✶ Average distance from the Sun: 2,794 miles (4,497 km)

✶ Mass (Earth=1): 17

✶ Spins on its axis in: 16.11 hours

✶ Circles around the Sun in: 165 years

✶ Number of moons: 8

The southern hemisphere of Triton seems to be covered with pinkish snow.

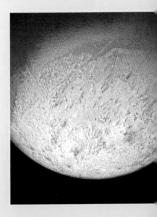

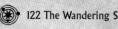

*Vague details on
Pluto's surface have
been spotted by the
Hubble Space Telescope.*

Pluto

The discovery of Neptune in 1846 did
not halt the search by astronomers for
new worlds. One of the keenest was the
ardent Martian enthusiast, Percival Lowell.
He worked out where he thought a new
ninth planet would be and set to work to
find it, using his observatory at Flagstaff,
Arizona. But he had no luck before he died
in 1916. Interest at the observatory in a
possible new planet waned until 1929,
when a young astronomer named Clyde Tombaugh
joined the staff and began the hunt again. By February
1930, he had found what he was looking for, a new
planet that came to be called Pluto.

Pluto is so tiny and so remote that it shows up
only as a star-like point even in the most powerful
telescopes. Currently, it is traveling through Ophiuchus
but it will slip into Sagittarius in 2007.

God of the Underworld

The Romans worshiped Pluto as god of the
Underworld, borrowing him from Greek mythology, in
which he was known as Hades. Hades was one of the
unfortunate children of Cronus and Rhea, whom his
father swallowed. But they were eventually liberated
by his brother, Zeus. A classic story tells of the
love of Hades for Persephone, the beautiful
daughter of Demeter, goddess of agriculture.
He carried her off to the Underworld
(see page 90).

A double planet?

Pluto is by far the smallest of the planets,
only about two-thirds the size of our
Moon. For most of the time it is the furthest

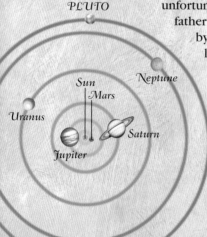

PLUTO

Sun

Mars

Neptune

Uranus

Jupiter

Saturn

planet. But for twenty years of the 248 years it spends circling the Sun, it slips inside Neptune's orbit, and that body becomes the furthest planet. The last time this happened was between 1979 and 1999. The reason for this is because Pluto's orbit is highly elliptical, or oval, which makes its distance from the Sun vary widely. Its orbit also takes it well above and well below the orbits of the other planets.

We know less about Pluto than about any other planet because it has not yet been visited by space probes. We do know it is a deep-frozen world, made up of rock and water ice, with frozen nitrogen and other gases covering its surface. It also has a moon circling around it, called Charon. This moon is half its size, prompting astronomers to talk about Pluto/Charon as a double planet. The latest research suggests that Pluto, Charon, and Neptune's moon Triton are part of a swarm of icy bodies that exist in deep space.

Pluto (Hades), pictured with Cerberus, the multi-headed watchdog of the underworld.

Factfile

✴ Diameter: 1,420 miles (2,284km)
✴ Average distance from the Sun: 3,670 million miles (5,900 million km)
✴ Mass (Earth=1): 0.002
✴ Spins on its axis in: 6.4 days
✴ Circles around the Sun in: 248 years
✴ Number of moons: 1

Below left: Clyde Tombaugh with the blink comparator he used to discover Pluto in 1930

Below right: Pluto and its moon Charon, discovered by James Christy in 1978.

Glossary

active galaxy A galaxy with an exceptional energy output.

asteroids Lumps of rock and metal that circle the Sun, mainly in a broad belt (band) between the orbits of Mars and Jupiter.

astrology A belief that the heavenly bodies somehow affect human lives.

astronomy The scientific study of the heavens and the heavenly bodies.

atmosphere The layer of gases that surround a heavenly body, in particular Earth's atmosphere.

atom The smallest unit of a chemical element, consisting of a nucleus with a swarm of electrons circling around it.

Big Bang A mighty explosion that scientists think created the universe some 15 billion years ago.

binary A two-star system in which the component stars are bound to each other by gravity and rotate around each other.

black hole A region in space with such powerful gravity that not even light can escape from it.

celestial sphere An imaginary dark sphere that early astronomers thought surrounded Earth.

circumpolar stars Stars close to the celestial poles, which remain visible every night.

cluster A group—of stars or galaxies.

comet A lump of icy matter that shines when it nears the Sun.

constellation A group of bright stars that form a pattern in the sky.

cosmos Another term for universe.

crater A pit in the surface of a planet, moon, or other solid heavenly body, made by a falling meteorite.

double star A star that is actually two stars that are close together or appear close together in the sky. (See binary).

eclipse What happens when one heavenly body moves in front of another, covering it up.

ecliptic The apparent path of the Sun around the celestial sphere each year.

equinoxes Times of the year when the lengths of daytime and nighttime are equal.

extraterrestrial Not of Earth; alien.

falling star A common name for a meteor.

galaxy A "star island" in space. Our galaxy is called the Milky Way.

gravity The force every piece of matter has on every other piece.

heavens The night sky.

interplanetary Between the planets.

interstellar Between the stars.

light year The distance light travels in a year—about 6 million million miles (10 million million kilometers).

lunar Relating to the Moon; hence lunar eclipse—an eclipse of the Moon.

magnitude The brightness of a star. Apparent brightness is a star's brightness as it appears to us. Absolute magnitude is the star's true brightness.

mare A sea, or plain on the Moon (plural maria).

meteor A streak in the night sky made when a speck of rock or metal from outer space burns up in Earth's atmosphere.

meteorite A piece of rock or metal from outer space that falls to the ground.

moon A common name for a satellite of a planet.

nebula A cloud of gas and dust in space.

nova A faint star that brightens suddenly and looks like a new star.

nuclear reaction A reaction involving the nuclei (centers) of atoms.

orbit The path one body follows when it circles around another in space.

phases The different shapes of the Moon during the month.

planet One of nine bodies that circle in space around the Sun.

precession A slight wobbling of Earth on its axis, which slowly changes the seasons, the time of the equinoxes, and the positions of the celestial poles.

probe A spacecraft that leaves Earth to travel to other heavenly bodies.

pulsar A rapidly rotating neutron star that gives off pulses of radiation.

quasar An object that looks like a star but is incredibly far away, and has the energy output of galaxies.

radio astronomy Astronomy that studies the radio waves the heavens give out.

reflector A telescope that uses mirrors to collect and focus starlight.

refractor A telescope that uses lenses to collect and focus starlight.

satellite A small body that orbits around a larger one; a moon. The usual term for an artificial Earth satellite.

shooting star A popular name for a meteor.

solar Relating to the Sun; hence solar eclipse, an eclipse of the Sun.

solar system The family of the Sun, including the planets, their moons, asteroids, and comets.

star A huge globe of intensely hot gas that gives off enormous energy, particularly as light and heat.

stellar Relating to stars.

supergiant The largest kind of star.

supernova The biggest explosion in the universe, when a massive star blows itself apart.

terrestrial Like, or relating to Earth.

universe Everything that exists—space and all the matter it contains, such as galaxies, stars, planets, gas, and dust.

variable A star that varies in brightness.

zodiac An imaginary band in the heavens through which the Sun and the planets appear to travel.

Greek Alphabet

α	alpha	ι	iota	ρ	rho
β	beta	κ	kappa	σ	sigma
γ	gamma	λ	lambda	τ	tau
δ	delta	μ	mu	υ	upsilon
ε	epsilon	ν	nu	φ	phi
ζ	zeta	ξ	xi	χ	chi
η	eta	o	omicron	ψ	psi
ϑ	theta	π	pi	ω	omega

Index

Page numbers in *italics* refer
to captions

Credits

Quarto would like to thank and acknowledge
the following for supplying pictures reproduced
in this book:

Key: b = Bottom, t = Top, c = Center, l = Left,
r = Right

AKG, Berlin: 16tl. **Anglo Australian
Observatory/Royal Observatory, Edinburgh**
(Photograph from UK Schmidt plates by David
Malin): 84tl. **Ann Ronan Picture Library**: 6cl,
8tl, 9tl, 10bl, 11tr, 12tl, 13br, 15br (NASA), 21br,
22tl & bl, 28tl, 31br, 38bl, 42tl, 50tl, 58tl, 61tr, 88tl,
94tl, 107br, 120cl. **The Art Archive**: 37br (Musée
du Louvre, Dagli Orti), 49tr (Galleria degli Uffizi,
Florence/Dagli Orti). **Bridgeman**: 56tl
(Christie's Images, London), 75tl (National
Museum of Scotland) **British Museum**: 9r.
Edimedia: 7tl, 26tl, 34l, 50bl. **Jason Ware**: 26bl.
Kopernik Space Age Center: 36tl. **NASA**: 24cl
(D Walter, South Carolina S. U. & P Scowen/
B Moore, Arizona S. U.), 24bl (ESA/M Romaniello,
ESO, Germany), 28bl (R O'Dell/K P Handron,
Rice U. Houston, Texas), 39br (CXC, SAO), 42bl
(Hubble Heritage Team AURA/STScI), 44bl, 47tr
(P. Seitzer, U. Michigan), 47br (E J Schreier, STScI),

54tl (N Walborn/J Maíz-Apellániz, STScI,
Baltimore, MD/R Barbá, La Plata Observatory,
Argentina), 54bl, 55br (Hubble Heritage Team
AURA/STScI), 59bl (ESA/ SOHO), 63br (SEDS),
64bl (JPL), 70bl (Hubble Heritage Team AURA/
STScI), 72bl (A Dupree, Harvard-Smithsonian
CfA/R Gilliland, STScI/ESA), 77br (SEDS), 79cr
(SEDS), 81br (A Caulet, ST-ECF/ESA), 82bl (SEDS),
85br (Hubble Heritage Team AURA/STScI), 87br
(SEDS), 94-95c (ESA/ SOHO), 96bl, 99br (JPL),
100tl, 102bl, bc & br, 103tr, 107tr, 108tl, 109tl &
br, 111tr (JPL/U. of Arizona), 112tl & bl, 113tl, tr
& b, 116bl, 117tr (Hubble Heritage Team AURA/
STScI), 117bl, 118-119c (JPL), 119tr & br (JPL)
120tl (JPL), 121br (JPL), 122tl (A Stern, Southwest
Research Institute/M Buie, Lowell Observatory/
ESA), 123br (Dr. R Albrecht, ESA/ESO). **Robin
Kerrod**: 20bl, 21tl, 32tl, 45tr, 52bl, 68tl, 74tl.
Royal Astronomical Society (Urania's Mirror):
67br, 78tl. **Science Photo Library** 123bl.

All other photographs and illustrations are the
copyright of Quarto. While every effort has been
made to credit contributors, we apologize should
there have been any omissions or errors.